MATERIALS MANAGEMENT
An Integrated Approach

MATERIALS MANAGEMENT
An Integrated Approach

P. GOPALAKRISHNAN
Administrative Staff College of India
Hyderabad

M. SUNDARESAN
Graduate School of Business
Columbia University
New York

PHI Learning Private Limited

Delhi - 110092

2026

*In fond memory of **Shri Asoke K. Ghosh** (October 1942 – February 2024), Founder Chairman and Managing Director of PHI Learning, whose vision endlessly inspires.*

The Legacy Continues....

Published by Pushpita Ghosh, PHI Learning Private Limited, Rimjhim House, 111, Patparganj Industrial Estate, Delhi-110092 and Printed by Gopsons Printers Pvt. Ltd., A-14, Sector-60, Noida, Gautambudh Nagar, Noida, Uttar Pradesh-201301.

₹675.00

MATERIALS MANAGEMENT—An Integrated Approach
by P. Gopalakrishnan and M. Sundaresan

ISBN-978-81-203-0027-9 (Print Book)
ISBN-978-93-90669-61-5 (e-Book)

The export rights of the book are vested solely with the publisher.

Contents

Preface

The main thrust of the book is on the Integrated Materials Management function suited to Indian conditions—where there is an acute paucity of authentic literature.

In the preparation of the book, we had the advantage of the feedback of the Indian situation, by organising executive development programmes for over a thousand senior professional materials managers at the Administrative Staff College of India. Our association with the Universities and the Professional bodies enabled us to get an overall perspective of the function. Moreover, the Consulting and Research assignments undertaken by us gave the book the crucial practical outlook.

The book is divided into four parts. Part I (pp. 1–51) covers the policy aspects of materials management including the concept of management by objectives. The various aspects relating to the purchasing management are dealt with in Part II (pp. 53–148). The managerial and operational functions of Warehousing and Stores are described in Part III (pp. 149–199). In the final part (pp. 201–260), Scientific Management principles including Inventory Control and Operational Research techniques are presented. A logical build-up has been ensured so as to enable the reader to gain a clear insight of the integrated materials management function. Wherever necessary examples and illustrations have been provided.

We feel that the book will be of value to the students of management at the postgraduate level in various universities and institutes of management. The practising materials managers in the industry will find this book refreshing and useful.

We wish to acknowledge the help rendered by a large number of colleagues and professional materials managers. In particular we would like to record our grateful thanks to Mr. S. Jambunathan, presently with New India Industries Limited, for his significant contribution, especially in the area of quantitative techniques. Our special thanks are also due to Miss Edna Pereira for her persevering task of typing the manuscript.

THE AUTHORS

Part I
POLICY ASPECTS

1

Integrated Materials Management

IMPORTANCE OF MATERIALS MANAGEMENT

The fast developing Indian economy has placed before the materials manager a tremendous challenge and responsibility. The task is really herculean when we recognise that more than Rs 15,000 crores worth of materials and components per annum go into the production channels, of which imports alone constitute about Rs 3,000 crores. The challenge he faces is all the more tough as the money tied up in inventory aggregates Rs 15,000 crores, of which obsolete materials lock up funds to the tune of Rs 2,500 crores.* The largest Central Purchasing Organisation in the country, namely, the Directorate General of Supplies and Disposal, buys over Rs 1,200 crores worth of materials per year.

In many organisations, materials form the largest single expenditure item. An analysis of the financial statements of a large number of private and public sector organisations indicates that materials account for nearly 60 per cent of the total expenditure. The information on the average materials expenditure for different manufacturing industries is given in Exhibit 1.1.

EXHIBIT 1.1
AVERAGE MATERIALS EXPENDITURE

Average expenditure of materials per cent	*Industry groups*
Above 65	Cotton yarn, earthmoving equipment, sugar, wool, jute, commercial vehicles, fabrication
60–65	Cotton textiles, bread
55–60	Engineering, non-ferrous
50–55	Shipbuilding, chemicals, tyre, machine tools, cement, electricity
45–50	Pharmaceuticals
40–45	Steel, newspaper, fertilizer, aircraft

Thus, the importance of materials management lies in the fact that any

*Figures taken from the speech delivered by Mr. C. Subramaniam, former Union Finance Minister, in his inaugural address to the Coimbatore branch of National Association of Materials Management, 11th December 1974, *The Hindu,* page 10.

significant contribution made by the materials manager in reducing materials cost will go a long way in improving the profitability and the rate of return on investment. Such increase in profitability, no doubt, can be effected by increasing sales. But with the increased competition and government restrictions on expansions this alternative is not easily achieved. Besides, some increase in the profitability can be achieved by concentrating on the materials cost which is typically a major rupee item for most organisations. In fact, as market pressure intensifies, organisations will be forced to cut down the costs and here the materials manager steps in to play his role.

Consider, for example, an organisation with an annual turnover of Rs 10 million, where materials account for about Rs 6 million and profits for Rs 1 million, and if a manager is given the work of improving the profitability, by, say, Rs 0.3 million, he has before him two ways of achieving the results. One is to increase the sales turnover by 30 per cent in a specific span of time, such that the overall profits will be Rs 1.3 million. The other alternative would be to reduce the materials cost by 5 per cent in a certain time dimension, so as to result in an increased profit margin of Rs 0.3 million. In other words, the result of the efforts in reducing the materials cost by 5 per cent is the same as the marketing efforts of increasing the sales by 30 per cent. Increasing the sales calls for a lot of additional marketing and advertising expertise, efforts, time and expenses, whereas if ways and means are diverted in saving a rupee in materials cost with the existing infrastructure facilities, it will go a long way in not only improving the profits but also in raising the corporate image of the company. In this context, the materials management experts like Dean Ammer, claim rightly that efforts for saving a rupee in materials cost is almost equal to the efforts made for additional sale of Rs 10. This is more true in a shortage economy, like that of India, and hence Indian management should devote pointed attention on materials function for reaping the full benefits.

Materials form an important part of the current assets in any organisation. The Return on Investment (ROI) depends a great deal on the manner of utilisation of materials. The relationship is represented below:

$$\text{ROI} = \frac{\text{Profit}}{\text{Sales}} \times \frac{\text{Sales}}{\text{Fixed assets} + \text{Current assets}}$$

Fixed assets constitute capital already sunk and the only scope for improving the return on investment lies therefore in the efficient management of materials which constitute the bulk of the current assets. Therefore, in this context, the control of materials assumes great importance.

NEED FOR INTEGRATED CONCEPT

Having recognised the importance of the materials management function, let us now see why an integrated approach is necessary. Various functions served by materials management include the materials planning, purchasing, receiving, stores, inventory control, scrap and surplus disposal. If some of the functions were to be separately handled, normally a conflict of interests occurs. Purchasing department, if allowed to operate independently, may take decisions which result in sub-optimisation. For example, under a separate set-up, the purchase department may treat discount as a very important factor and buy large quantities to avail of the discount without taking into account its impact on the warehousing and carrying costs. In other words, we need to balance the conflicting objectives from a total organisation viewpoint so as to achieve optimum results for the organisation as a whole. An expansion, for example, will require planning for the increased requirements, developing new sources, revision in inventory levels, apart from the increased load in receipt of materials, inspection and storing.

In an integrated set-up, the materials manager who is responsible for all such interrelated functions, is in a position to exercise control and coordinate with an overview that ensures proper balance of the conflicting objectives of the individual functions. Integration also helps in the rapid transfer of data, through effective and informal communication channels. This is crucial as the materials management function usually involves handling a vast amount of data. Therefore, integrating the various functions ensures that message channels are shortened and the various functions identify themselves to a common materials management department which, in turn, results in greater coordination and better control.

DEFINITION AND SCOPE

We can define materials management as the function responsible for the coordination of planning, sourcing, purchasing, moving, storing and controlling materials in an optimum manner so as to provide a pre-decided service to the customer at a minimum cost.

From the definition it is clear that the scope of materials management is vast. We can broadly identify the following functions:

Materials Planning and Control

Based on the sales forecast and production plans, the materials planning and control is done. This involves estimating the individual requirements of parts, preparing materials budget, forecasting the levels of inventories, scheduling the orders and monitoring the performance in relation to production and sales.

Purchasing

This includes selection of sources of supply, finalisation of terms of purchase, placement of purchase orders, follow-up, maintenance of smooth relations with suppliers, approval of payments to suppliers, evaluating and rating suppliers.

Stores and Inventory Control

This involves physical control of materials, preservation of stores, minimisation of obsolescence and damage through timely disposal and efficient handling, maintenance of stores records, proper location and stocking. Stores is also responsible for the physical verification of stocks and reconciling them with book figures. The inventory control covers aspects such as setting inventory levels, ABC analysis, fixing economical ordering quantities, setting safety stock levels, lead time analysis and reporting.

ADVANTAGES IN INTEGRATED MATERIALS MANAGEMENT CONCEPT

Organisations which have gone in a big way for the integrated materials management usually enjoy the following advantages:

Better Accountability

Through centralisation of authority and responsibility for all aspects of materials function, a clear cut accountability is established. Various user departments can direct their problems with regard to materials to one central point so that action can be taken immediately. This helps in evaluating the performance of materials management in an objective manner.

Better Coordination

When a central materials manager is responsible for all functions, the departments under the materials manager create an identity which is common. This results in better support and cooperation in the accomplishment of the materials function. The user departments also find that they have to approach one department for discussing and solving their materials problems. This creates an atmosphere of trust and generally better relations between the user departments and the materials management department.

Better Performance

As all the interrelated functions are integrated organisationally, greater speed and accuracy results in communication. Need for materials is promptly brought to notice by materials planning. Purchase department is fed with stock levels and order status by stores and inventory control departments. All this calls for judicious decisions leading to lower costs, better inventory turnover, reduced stock-outs, reduced lead times and a general reduction in paper work.

Adaptability to EDP

The centralisation of the materials function has made it possible to design data processing systems. All information with regard to materials function is centralised under the integrated materials management function. This has facilitated the collection, collation and analysis of data, leading to better decisions. Advanced and efficient electronic data processing systems can be economically introduced under an integrated set-up.

Miscellaneous Advantages

Under a central materials manager, a team spirit is inculcated. This results in better morale and cooperation. The opportunities for growth and development are better in an integrated set-up. An individual under such a set-up is not confined to any one function alone, and he gets, over a period, exposed to broader aspects of the materials function.

FUTURE

Many organisations in India are at present wedded to the existing system. Only very recently, the integrated materials management concept has gained greater acceptance. Although organisations may not be adopting a totally integrated approach, there is a definite tendency to move towards the integrated approach. This has also necessitated professional development of managers so that they can fulfil the requirements of an integrated materials management function which demands an ability to bring together conflicting and yet interrelated functions. The economic pressures, such as inflation and credit squeeze, will place exacting demands on the materials manager in future. This will sharpen the professional acumen of the materials manager.

2

Corporate Policy and Materials Management

GENERAL CORPORATE POLICY

Corporate policies are laid down in order to achieve the corporate goals that the organisation has set for itself. The process is explained in Exhibit 2.1. Many

EXHIBIT 2.1

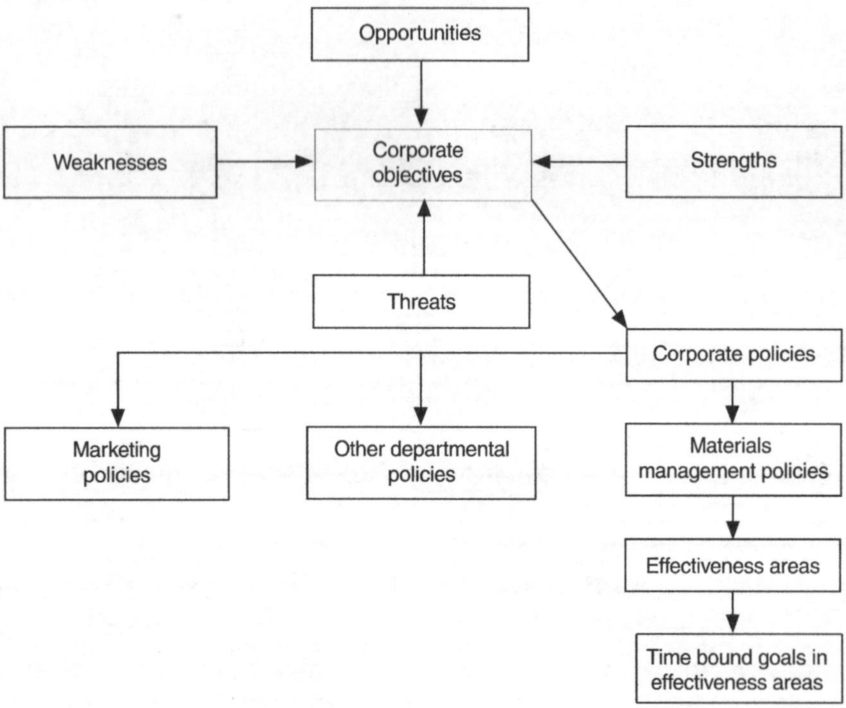

organisations have multiple goals. Needless to add, these goals must be measurable, result-oriented, specific, realistic and attainable in a time span. Some of the important goals are to

1. achieve a specific net profit,
2. achieve a specific sales target,
3. increase the capacity utilisation,
4. obtain a fixed return on investment,
5. maximise shareholders' earnings, and
6. improve the corporate image.

Based on the corporate goals, broad policies are laid down in order to attain them within a definite time span. The top management therefore expects that each department must work towards the organisational objectives. In fact, they expect individual departments to directly contribute to the objectives through well defined policies and plans of action. For example, the top management may expect the materials department to cut down inventories and yet maintain a pre-decided service level and thereby contribute to the overall profitability. This is logical as the materials management department, probably, is responsible in many organisations for nearly 70 per cent of the cash outflows.

The Concept of MBO

Management by Objectives (MBO) is a tool which is increasingly used by many organisations. W.J. Reddin defines MBO* as "the establishment of effectiveness areas and effectiveness standards for managerial positions and the periodic conversion of these into measurable time-bound objectives linked vertically and horizontally and with future planning".

In organisations, effectiveness areas will have to be identified in Production, Sales, Finance, Materials and Personnel functions; objectives should be set linking them to the overall organisational objectives, and then control should be exercised. In materials management, the effectiveness areas may be identified as follows:

1. Inventory levels
2. Inventory ratio
3. Rejection percentage per supplier/per item
4. Scrap
5. Unit cost.

After identifying the effectiveness areas, the specific output requirements and measurement criteria will have to be established. They must then be converted to time-bound targets. Thus, the objectives and policies of materials management are derived from the corporate goals and policies. This is illustrated in Exhibit 2.1.

A clearly defined materials management policy enables the spelling out of authority and responsibility. It also helps to identify the prerogatives of the materials management to avoid confusion that may arise in inter-departmental activities. Within the materials management area, it facilitates uniform and

*Reddin, W.J., *Effective Management by Objectives,* McGraw-Hill, page 12.

consistent action in various activities. Procedures and systems get standardised, leading to greater overall efficiency.

SCOPE

Policy should be spelled out in the following areas:

Organisation

Acceptance of integrated materials management itself is a major policy decision. The internal structure, delegation of authority, interface relations, coordination of the entire materials management function, identifying training needs, and developing a policy manual are some of the important areas which come under this category.

Purchasing

Policy must be clearly laid down with regard to source development, quotations, receiving salesmen, negotiations, price, quality, plant visits, personal buying, reciprocity, capital equipment buying and make or buy decisions. These are very important areas where purchase department must be given exclusive powers. If other departments like production, maintenance and engineering require some technical details about new materials from the suppliers, then they must inform the purchase department about correspondences that they plan to have with suppliers in this regard.

Stores

Procedures must be established for receipt, inspection, handling, storing, issue, stock verification, surplus and obsolete material disposal so that the inventory control levels are kept at a desired level. This will also ensure proper accounting, safety, minimum lead time, and eliminate pilferage.

Policy Manual

In view of the complex nature of the materials management function and its important contribution to the overall company objectives, many organisations have found it desirable to compile a policy manual covering all aspects of materials management. The manual is generally prepared with a note from the Chairman of the company emphasising its importance. Such a policy manual will be circulated to all divisions up to the middle management level so as to make certain that policies and procedures are consistently followed leading to the achievement of the corporate objectives. The policy manual familiarises the materials management department with the procedures and guidelines to be observed. Periodical revision of the manual needs to be undertaken to keep it realistic and responsive. Typically, the manual will have the scope which we earlier discussed, and will also stipulate many procedural guidelines. A sample policy manual is shown in Exhibit 2.2.

EXHIBIT 2.2
POLICY MANUAL

(a) *XYZ Company:*

Policy Manual (Materials Management)

I have pleasure in sending you a copy of the policy manual of materials management. The policies and procedures laid down in this manual are very essential in achieving the objectives of our company. This manual will help you in clearly understanding the objectives, policies and responsibilities of the materials management department.

Sd/-
Chairman
XYZ Company

(b) *Contents:* (The points outlined here are only illustrative and not exhaustive.)

1. Materials Management is the only department which will serve as the point of contact between suppliers and the company.

2. Any department which may wish to contact a supplier directly must do so after clearance from materials management and must keep it informed of all transactions.

3. Source development is the exclusive responsibility of materials management department.

4. Other company's salesmen can be received only by the materials management department. Other departments may receive them after clearance from materials management.

5. Materials Management can question the quality, quantity and price of the materials asked for by the user department, in the best interests of the company.

6. The responsibility for analysis of requisitions, sending out enquiries, analysis of quotations, placement of purchase orders, expediting and controlling inventory levels, lie entirely with materials management.

7. The practices of reciprocity, personnel buying, etc., will be done by materials management under top management decisions. (It may be sometimes explicitly stated.)

8. Materials Management shall keep the concerned departments informed of the delivery schedules, accounts payable, receipts, etc., so as to ensure smooth operations.

9. Market price levels, new product developments, etc., will be communicated to the concerned departments by the materials management.

10. In order to effectively perform its function, materials management department will be assisted by the finance department in prompt clearance of supplier bills and in insuring the stock-in-transit and stores.

11. Marketing and production departments shall keep the materials management department informed of future plans so that expeditious action can be taken in controlling inventory levels.

12. The materials management department will assist top management in capital equipment buying through locating sources and analysing the merits of various alternatives.

The policy aspects when laid down in a systematic manner will ensure that the operating personnel are clear about their responsibilities and the procedures they have to follow so that the overall organisational goals are easily achieved.

MAKE OR BUY

Whether to make an item within the user's plant or to buy it from outside sources is a major policy decision. Normally decisions in this aspect are taken at top level.

The role of materials management here depends upon the position of the head of the materials management division in the organisation. If he is a member of the top management team, then a key role is played by him as he is best equipped to deal with the problem. Otherwise he has to play a supporting role by providing relevant data so that the top management can take a scientific decision. A great deal of coordination is required between materials management and production departments in collection, collation and analysis of data, such as quality requirements, quantity requirements, existing production facilities, balance equipment requirements and so on. Few organisations have formed committees to analyse and recommend proposals. Make or buy decisions arise due to a variety of reasons. Mostly such situations arise due to consistent failures on the part of suppliers in the supply of items. The management then decides to make the item. Sometimes availability of items from outside suppliers may not be adequate owing to stiff competition from many end-users. This also may lead to making the item in one's own plant.

Factors that are to be carefully analysed in arriving at make or buy decision include the following:

1. Quality requirements
2. Quantity requirements
3. Cost aspects.

All the above factors have to be considered in depth before a decision is taken.

QUALITY REQUIREMENTS

If the suppliers consistently fail to meet the quality requirements, then the firm may decide to make the item. Firstly, it is essential to find out whether such exacting quality requirements are necessary at all. It may mean specialised procedures, tooling, skilled labour and rigid inspection standards. Another important aspect is whether the buyer will be able to make the item to such quality standards when established vendors are unable to do so. A vendor enjoys the benefit of specialisation and probably the item in question is his main product line. Buyers also have the option of changing the supplier and hence there is a lot of flexibility.

QUANTITY REQUIREMENTS

When the quantum is small, such as specialised jigs and fixtures, it may be worthwhile to make them especially if the existing facilities and the little amount of sub-contracting is adequate for such manufactures. It is not a paying proposition for the supplier to accept such small jobs. Sometimes demand for items may not be certain and it may be advantageous to make it using a flexible production schedule. When the requirements are very large, generally buying would be desirable.

COST ASPECTS

Advantages and disadvantages will have to be quantified and a cost analysis has to be done. If existing capacity can be used then fixed costs do not normally enter the analysis. However, when fresh capacity is to be added, additional fixed costs will have to be incurred and have to be taken into account. In utilising the existing capacity for making an item, care should be taken to see that the projected sales requirements of the finished products are not affected for want of capacity.

Factors such as the cost of transportation of raw materials vis-a-vis finished item, economies of scale and the cost of capital must be given a thorough analysis in arriving at a make or buy decision. A detailed note on marginal costing is given in Exhibit 2.3.

EXHIBIT 2.3

Marginal costing is very useful with regard to decisions involving make/buy. This technique particularly avoids the difficulties experienced in allocating fixed costs in the conventional manner to a number of processes or products. The conventional method of allocating fixed costs amongst different products does not delineate the contribution made by the individual products to the profit. Let us examine this by means of an example.

Conventional Analysis

	Rs	Rs
Sales revenue	—	4,000
Variable expenses:		
Material costs	1,300	
Wages and expenses	1,240	
Fixed expenses	1,400	3,940
	Profit	60

Marginal Analysis

	Rs	Rs
Sales revenue	—	4,000
Variable expenses:		
Material costs	1,300	
Wages and expenses	1,240	
		2,540
Recovered fixed costs	Costs	1,460

By selling this particular article, the company has recovered Rs 1,460 as a contribution to total fixed costs. Each such article would then contribute to the recovery of the total fixed costs, and when the entire fixed costs are recovered, each additional article would produce a profit of Rs 1,460. If more than one product is involved, then the contributory margin can be worked out for each product as the difference between the price and the variable cost. Hence, each product would make a contribution to the recovery of fixed costs, which are regarded as a whole with no attempt made to allocate them. The products with the largest contribution margin are preferred for "making" in the organisation.

Other than the three factors discussed above, factors such as flexibility, technical know-how, labour content, usage level, etc., also influence the top management's policy in make or buy decisions.

3

Organisation and Control

The inter-related functions of materials management are normally looked after by individuals. As the activities expand, the functions of each individual become more and more specialised. The integrated materials management concept requires central coordination of all these inter-related activities. Therefore the internal structuring of the various functions as well as the relationship of the materials management division with the other divisions—technical, finance and marketing—in the overall organisation becomes critical. The materials management function ought to be headed by a competent professional who must be a member of the top management team as managing materials is a critical function.

MATERIALS MANAGEMENT IN THE OVERALL COMPANY ORGANISATION

We have seen the need for integrated materials management and the advantages that go with it. It was also recognised that materials management is a top management function. Some organisations in India have a director in the board exclusively in charge of materials and even small organisations are no exception to the order. A typical organisation chart is shown in Exhibit 3.1.

EXHIBIT 3.1
TOTAL ORGANISATIONAL PLAN

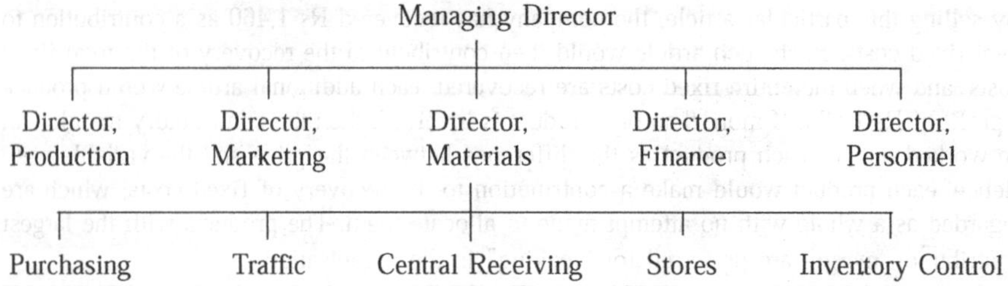

Depending upon the company's operations and its environment, many variations of the organisation plan are feasible.

ORGANISING MATERIALS MANAGEMENT

Once the relationship of materials management with regard to other departments in the overall organisational plan is finalised, the next step is to finalise the internal structuring of the department. The following alternatives are available.

ORGANISATION BASED ON COMMODITIES

Under this system of organisation, items are classified according to their nature such as raw materials, bought out components, spares, imported items, finished goods and so on, and they are assigned to individuals. Depending upon the importance of the commodity group in respect of the operations of the company, workload will vary between the groups and this forms the basis for determining the staff in each commodity group. For example, an automobile manufacturing firm may have commodity groups for forgings, castings, pressed parts, raw materials, finished bought out items, and civil engineering materials. This arrangement is shown in Exhibit 3.2. The advantage in such a system is that there is no wasteful duplication of efforts as each commodity is separately handled. The group is intimately in touch with its respective commodity market and hence it becomes specialised. Bulk buying and standardisation are facilitated in each commodity group.

EXHIBIT 3.2
COMMODITY GROUPING: ORGANISATION CHART

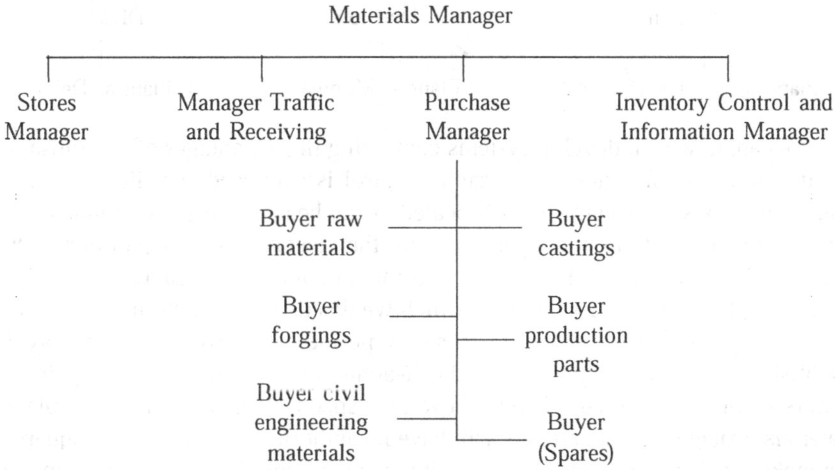

ORGANISATION BASED ON LOCATION

When an organisation has several plants located in different parts of the country there are two alternatives. One is to have a centralised organisation located at the headquarters. The other is to have a decentralised materials management set up in each location. There are several reasons favouring the latter alternative. Firstly, when distances between the plants are significant, the materials management function is impaired when controlled from headquarters, both in terms of cost and time. Secondly, the plant's product lines play an important role. Each plant may require many unique materials and a materials management department located at the plant will be in a much better coordination with the plant's production, finance and marketing departments.

We can likewise list the advantages of a centralised materials management department for multi-plant operations. The most important benefit is that combining the requirements of plants and buying in bulk leads to substantial reduction in cost. In a centralised set-up, inter-plant transfer of materials in an emergency is possible. Also, surplus materials in one plant can be utilised in meeting the requirements of another plant. Since there is only one department dealing with various sources of supply, a high order of skill can be expected. An example of decentralised set-up is shown in Exhibit 3.3.

EXHIBIT 3.3
MULTIPLANT OPERATIONS

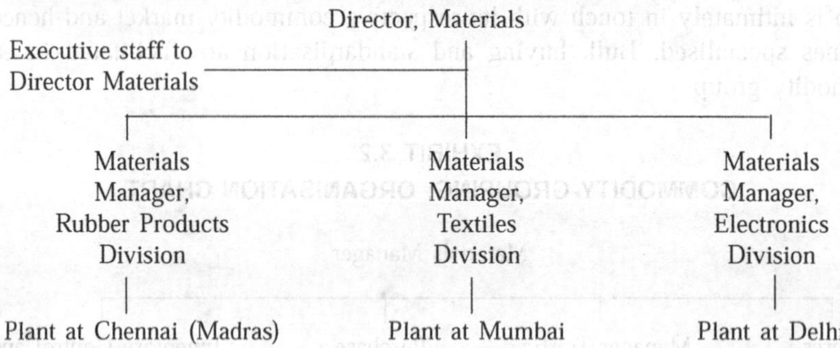

We can, however, develop systems combining the advantages of centralisation and decentralisation. Under this system, control is exercised at policy level by a central materials management staff located at the headquarters. Continuous feedback is maintained on the operations of the decentralised departments. For example, the headquarters may lay down common policies, procedures and systems for all the plants. The policy manual which we discussed at length in the previous chapter may be used for this purpose. A periodic reporting system may be established between the plants and the headquarter. A system of delegation of powers could be installed whereby beyond a specific financial limit the plant's materials management department will have no authority and only the headquarters can make the decision. Common usage items where bulk buying could mean a lot of saving may be exclusively handled by the headquarters. Organisations having multi-plant operations would benefit by adopting a judicious blend of decentralised set-up with centralised control.

ORGANISATION BASED ON FUNCTION

Here the materials management department is structured on the basis of functions such as stores, transport, receiving, purchase and so on. Thus, stores of different manufacturing divisions will be under one individual who will report to the materials manager. All purchasing activities will be again under one individual who will report to the materials manager. This system which permits specialisation in each function is now followed by many organisations in India.

SPECIAL REQUIREMENTS

For organisations undertaking turn-key projects certain special aspects need to be considered. Erection of steel plants, commissioning of refineries, laying pipelines are some of the examples in this category. The important aspects which must be considered are:

1. The project schedule will require timely procurement of materials
2. Storage at site where conventional stores management aspects are not relevant
3. Flexibility of the set-up to spot interchangeable materials and equipment between projects to avoid delay and idling of equipment
4. Ability to obtain or forecast costs which will assist in costing the entire project.

This has become one of the major activities; following each five year plan more and more projects are being commissioned. Sometimes a division-wise materials management function is organised in multi-divisional organisations. This has one advantage in that the materials management department is quick to grapple with the problems and needs and this facilitates better coordination.

INTER-DEPARTMENTAL RELATIONSHIPS

Broadly, the materials management will have to work in close coordination with production, marketing, and finance departments. Only an atmosphere of mutual trust will ensure that these departments will work towards the total organisational objectives. For instance, production will have to keep the materials management department informed about its plans and schedules so that materials requirements can be planned in advance. Even the adjustments in sales forecasting, or the changes in schedules must be duly conveyed to materials management. In the same way, the materials management department must keep the production department informed about the lists of suppliers, availability of new materials and anticipated delays so that re-scheduling of production could be done and costly stock-outs avoided. The expertise of the materials department could be profitably used by the production department in purchasing capital equipments.

The finance department also has to work in close coordination with materials in anticipating funds requirements, payment of bills to suppliers, insurance and so on. Such close coordination will ensure prompt payment of bills and improve relationships with the suppliers.

Marketing department will have to give advance information on forecasts and special requirements so that planning can be done effectively. Materials Management through efficient operations can keep the prices at a competitive level and thus help the marketing department in its operations.

In many organisations formal committees consisting of executives drawn from Marketing, Production, Finance and Materials Management departments are formed to finalise (periodic) annual sales targets, production programmes, total budget, materials budget, etc. Formal organisational arrangements help in furthering inter-departmental relationships.

SPAN OF CONTROL AND DELEGATION

Principles of sound organisation should be followed in establishing span of control and delegation of powers. Generally speaking, executives at the top of an organisation will have fewer persons reporting to them than a materials engineer who at his level may have many persons reporting to him. Powers, especially those of expenditures, should be clearly defined so that smooth working is possible.

4

Materials Research

NEED AND IMPORTANCE

The twenty-five years of industrialisation of the Indian economy has made the materials manager shoulder heavy responsibilities. Naturally, his horizon of work has been widening and he is expected to be in constant touch with supply sources, market trends, and price-levels. What is more, the import of materials is a continuous drain on our foreign exchange reserve. It is thus imperative to develop alternative sources for materials so as to reduce imports. All this and the sheer value of the materials purchased every year point to the need for materials research on a short-term as well as long-term basis. The materials manager today has to consider the implications of the following factors:

1. The company's long range requirements of material.
2. The likely sources of supply and their capacities over the next 10 years. Anticipation of shortages in a particular sector.
3. The technological changes expected in the markets from which materials are purchased. The major breakthroughs, if any.
4. The likely price trends.
5. The expected changes in the Government policy in the related areas.

Definitely these aspects are critical in terms of the long-term viability of the materials management function. Systematic materials research will enable us to understand the implications of the factors listed above.

DEFINITION AND SCOPE

Materials Research can be defined as a systematic, formal and continuous analysis of all factors affecting the materials management function. The aim of the research is to ensure optimal decisions, particularly for important items and key raw materials. Thus, the scope of Materials Research is vast and is listed below:

1. Economic Analysis

In a planned economy such as India, the economic policies have constantly influenced the major decisions taken by various organisations. Factors, such as GNP, planned growth rate, business cycles, inflation, channelisation and other factors can seriously affect the availability of materials and consequently the price levels. Hence a systematic analysis is necessary in this area.

2. Market Analysis

The corporate objectives and requirements must be estimated over the next 5 to 10 years. The markets should then be analysed for supply trends, technological innovations, substitutes, evolution of new material, and standards. 'Critical' materials from the Company's point of view, in terms of anticipated shortage, can then be isolated for top management's attention. Materials have played a very important role in the progress of technology, ever since steel became the major material of engineering, aluminium became the commercial metal, Hyatl invented celluloid and Bakeland developed bakelite. Now materials are developed almost at an exponential rate. A detailed study has been done by Henry R. Clansers in his paper entitled "Thinking Ahead the Materials Age"* and by Laurence H. Van Vlack in his paper entitled "The Two Major Trends in Materials Education."** Rubber and plastic materials, which were relatively unknown two decades ago, have increased thousandfold and the growth still continues. Ceramic materials have also developed rapidly. In the field of metals there are atleast 20,000 different alloys (see Exhibit 4.1). Through organised materials research, a company can identify the correct material to suit its product design. This, in turn, is very important for the successful operation of the company. Delphi techniques, dealt with in detail in a later chapter, can be employed in market analysis where quantitative information is lacking.

3. Supplier Analysis

A checklist of the present and future suppliers is the first step. This should be followed by the analysis of their financial condition, their expansion plans, and additions to their product lines. An estimate of the share that the company can hope to get from their expanded capacity and new product lines can then be worked out. This will also highlight areas where new sources have to be developed. On a continuous basis the facilities of the suppliers should be inspected and critically evaluated. This analysis is necessary if sudden supply failures are to be avoided. It helps in planning for future operations also.

4. Transportation Analysis

Transportation cost forms a good percentage—about 20 per cent—of the cost of material purchase in any organisation. Analysis of the location of suppliers must be made in relation to the transportation cost. The recent energy crisis has considerably affected the cost of transportation, making it necessary to take a fresh look at the transportation mix, namely, the proportion of goods moved by

* *Harvard Business Review*, May-June 1961.

** *Materials in Design Engineering*, U.S., Sept. 1960.

trucks, railways, ships and aeroplanes. Many new developments are taking place. Railways, for example, have introduced Container Service, Quick Transit Service, Freight Forwarder Scheme, Mobile Booking Service, Parcel Express, etc. It pays for the organisation to avail of such facilities wherever suitable. Applicability of techniques, such as Linear Programming can also be explored in order to minimise transportation costs.

EXHIBIT 4.1

INCREASE IN VARIETIES OF ENGINEERING MATERIAL SINCE 1900

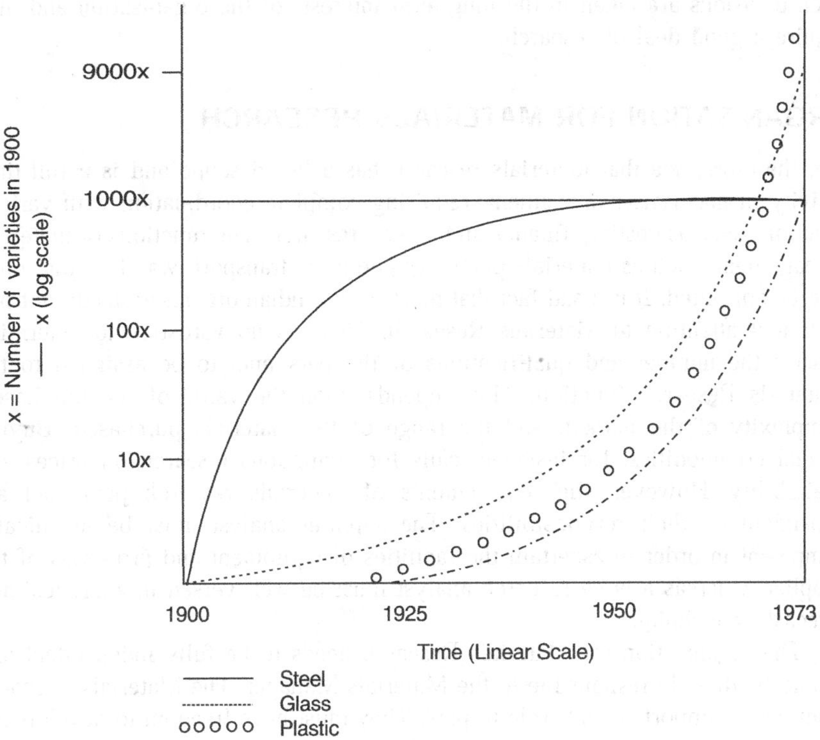

5. Price Analysis

Analysis of price increases made by the suppliers should be done on a regular basis and their impact should be related to the cost of finished products. For this purpose it is essential to have a detailed analysis of the cost structure of the raw material. For similar groups of materials such price analysis should be made and compared for various vendors. The causes should be identified so that various alternatives, such as use of alternate materials, change of suppliers, and so on, can be worked out.

6. Purchase Analysis

Advantages and disadvantages in purchasing from manufacturers vis-a-vis dealers should be analysed in the light of company's operations. Cost benefit analysis must be done in vertical integration of critical raw materials and components. An accepted policy in a few organisations is the concept of vertical integration, i.e., the process in which the company decides to manufacture some of the critical raw materials and components. For this purpose, detailed cost-benefit analysis must be done.

A steel manufacturer, for example, may decide to acquire a coal mine as it is critical to the operations, particularly when the supply of coal is not satisfactory. An automobile manufacturer may decide to have captive foundry and forge units. Such decisions are taken in the long-term interests of the organisation and they require a good deal of research.

ORGANISATION FOR MATERIALS RESEARCH

We, therefore, see that materials research has a broad scope and is a full-time activity in materials management, requiring complete coordination with various functions such as costing, finance and market research. The functions of materials management, such as materials planning, purchase, transport, warehousing, must also be consulted. It is a sad fact that most of the Indian organisations do not pay sufficient attention to Materials Research. There is no yardstick to determine exactly the number and qualifications of the personnel to be assigned to the Materials Research function. This depends upon the value of the purchases, complexity of the market, and the range of the materials purchased. Buying special commodities, for instance, calls for continuous research on prices and availability. However, the requirements of materials research personnel are dependent on their responsibilities. The supplier analyst must be technically competent in order to ascertain the facilities of equipment and processes of the supplier, whereas a price research analyst must be well versed in analytical and forecasting techniques.

The organisation for Materials Research needs to be fully independent and should be directly responsible to the Materials Manager. The Materials Research team has an important staff role to play. They must have freedom to develop and execute new research projects. Most of the organisations have a fairly established marketing research team and their expertise can be very helpful to the materials research team.

TECHNIQUES AND REPORTING

1. SWOT Analysis

The SWOT analysis—Strengths, Weaknesses, Opportunities and Threats analysis—when carried out in the Materials Research function at Company, Industry, National and International level helps to formulate strategies for optimal decisions. The information which the executive in charge of materials research function requires is normally available in India in the list of published literature given in Exhibit 4.2.

EXHIBIT 4.2
TITLE AND FREQUENCY

A. General

1. Statistical Abstract India (A)
2. Basic Statistics Relating to the Indian Economy (A)
3. Monthly Abstracts of Statistics (M)
4. Reserve Bank of India Bulletin (M) (Current Statistics)
5. Weekly Statistical Supplement to the Reserve Bank of India Bulletin

B. Agriculture

6. Estimates of Area and Production of Principal Groups (A)
7. Estimates of Area and Production of Principal Groups (A)—Detailed tables in 2 parts
8. Indian Agricultural Statistics
9. Indian Agricultural Statistics (Detailed tables)
10. Agricultural Situation in India (M)

C. Balance of Payment

11. Reserve Bank of India Bulletin (M) (Tables 41 and 42)
12. Report on Currency and Finance (A)

D. Foreign Trade

13. Monthly Statistics of the Foreign Trade in India, Vol. 1, Exports & Re-exports, Vol. 2, Imports

E. Industrial Statistics

14. Annual Survey of Industries (Census Sector), 10 Vols.
15. Annual Survey of Industries (Provisional Results) General Review (A)
16. Monthly Statistics of the Production of Selected Industries of India (irregular)
17. Monthly Bulletin showing Production of Selected Industries of India and the Index of Industrial Production (M)
18. Indian Minerals Yearbook (A)

F. Labour Statistics

19. Indian Labour Statistics (A)

G. National Accounts

20. Estimates of National Product (A)

H. National Sample Survey

I. Prices

21. Index Numbers of Wholesale Prices in India (A)
22. [Weekly Statistical Supplement to the Reserve Bank of India Bulletin gives group and sub-groupwise data with a time-lag of about three weeks]

J. Transportation

23. Basic Road Statistics (A)
24. Indian Transport Statistics (A)
25. Monthly Railway Statistics (M)
26. Review of Public Sector Road Transport (A)

A = Annual; M = Monthly.

2. Selective Approach

The ABC analysis approach can be successfully applied in conducting materials research. For materials which account for 80 per cent of the rupee value of purchases a formalised materials research has to be undertaken. It should be recognised that if a selective approach is not taken, then the research team may be flooded with projects which do not result in significant pay-offs. Therefore, the first step is to identify projects which have a significant pay-off in a short term. Some of the research projects which do have a significant pay-off may be beyond the capabilities of the research staff in the organisation. As we have observed earlier, very few organisations have an adequate research team. It is therefore necessary that projects of this nature are assigned to outside agencies specialising in this function.

For such projects which can be handled by the company's research organisation, the research staff should work in close coordination right from the design stage until the product is rolled out. All aspects discussed under the scope of materials research should be covered. Rating technique is usually adopted in directing the priorities of the research effort. Materials are rated on the factors such as (1) quality, (2) availability/location, (3) price/cost details, (4) product profitability/value, and (5) purchase aspects.

Depending upon the rating of the various factors, research activities are focussed on one or more of the elements such as design changes, use of alternate materials/substitutes, combining shipments, etc.

Reporting on the activities of materials research is the last function but it is as important as the other functions of research. It is essential that research projects undertaken are presented in framework which highlights the findings to top management. Functional reporting is very effective and could be presented covering the following aspects:

1. Company's requirements of materials for next 2 years, 5 years and 10 years
2. Anticipated prices of A items over the next 2 years, 5 years and 10 years
3. Source development needs to meet shortages over next 2 years, 5 years and 10 years
4. Funds requirements for materials over next 2 years, 5 years and 10 years
5. Trend of material cost over the next 2 years, 5 years and 10 years.

In view of the importance of materials research and benefits accrued from it, organisation should allocate a portion of the research expenses towards materials research.

5

Materials Planning and Budgeting

IMPORTANCE

The quintessence of any Manager's job is planning and control. Hence the materials planning and budgeting function is given a prominent place in the integrated materials management set-up. This is so because, planning for materials and working out a realistic budget not only help motivate people but also serve as a control device. Planning is done at all levels of the organisation. This is illustrated in Exhibit 5.1.

EXHIBIT 5.1

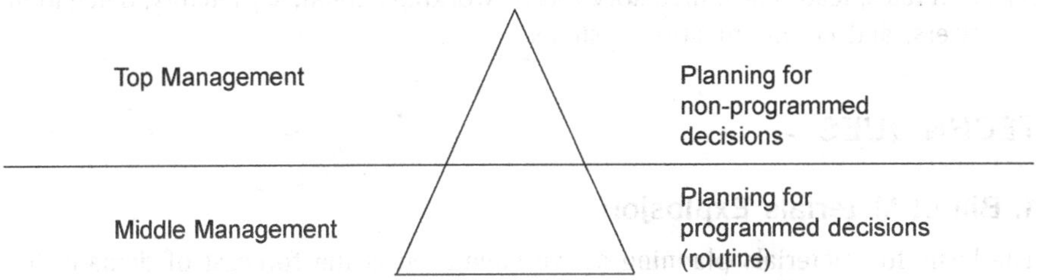

In the context of materials management, planning has to be done for highly non-programmed decisions such as import policy, foreign exchange availability and credit squeeze. In a similar manner, planning has also to be done for highly programmed decisions such as hypothecating inventory for working capital, working out delivery schedules, etc. The relationship between materials planning and other major functions are shown in Exhibit 5.2.

EXHIBIT 5.2

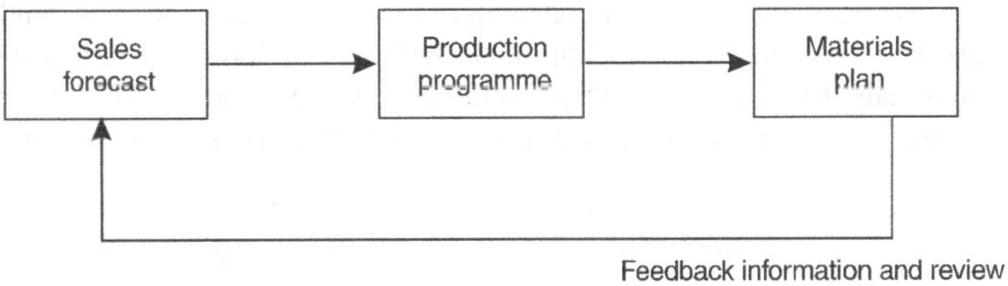

DEFINITION

Materials planning is the scientific way of determining the requirements of raw materials, components, spares and other items that go into meeting production needs within the economic investment policies. Thus, by definition, it follows that the materials planning function is a sub-system in the overall planning activity. This, too, is evidenced by Exhibit 5.2. The factors which affect materials planning can be classified in the following two categories:

1. ***Macro Factors.*** Some of the macro factors which affect materials planning are price trends, business cycles, government import policy, credit policy, etc. The recent credit squeeze followed by the interim report submitted by the study group to frame guidelines for follow-up of bank credit is an excellent case in point. This study group headed by Mr. P.L. Tandon has now given guidelines for inventory levels of raw materials, stocks in process, finished goods as well as for receivables and bills purchased and discounted.* Industry-wise details are furnished in Exhibit 5.3. No materials planner can afford to ignore these guidelines, as banks extend credit only as per these guidelines. Such macro level policy changes are bound to take place in future as well and should be considered while evolving the materials plan.

2. ***Micro Factors.*** Materials planner also has to take into account various factors at micro level. They include corporate objectives, plant capacity utilisation, rejection rates, lead times, inventory levels, working capital, seasonality, delegation of powers, and communication systems.

TECHNIQUES

1. Bill of Materials Explosion

The basis for materials planning as we have seen is the forecast of demand for the end-products. Under the chapter on "Price Forecasting", we will see various forecasting techniques, such as moving averages method, exponential smoothing, and time series. These techniques are equally applicable to demand forecasting as well.

Once the demand forecast is made, it is possible to go through the exercise of materials planning. Requirements of various materials are arrived at from the demand forecast, using bill of materials, through explosion charts. Computers can be very effectively used for "exploding" bill of materials with demand forecasts. Bill of materials is nothing but a document which shows for a given component the list of materials required, unit consumption, and location code. The condition of supply such as bought out or shop made will also be indicated. For a sample bill of materials, the reader is referred to the chapter on "Computers in Materials Management". An explosion chart is just a series of bills of materials grouped

---

* Report of the *Study Group to Frame Guidelines for Follow-up of Bank Credit,* Reserve Bank of India, Bombay, 1975.

EXHIBIT 5.3
*SUGGESTED NORMS FOR INVENTORY AND RECEIVABLES

Industry	*Raw materials (including stores and other items used in the process of manufacture)*	*Stocks in process*	*Finished goods*	*Receivables and bills purchased and discounted*
1	2	3	4	5
(i) Cotton and synthetic textiles	Cotton 2 (Bombay and Ahmedabad areas) 3 (Eastern areas-Bihar, Orissa, West Bengal and Assam) $2^1/_2$ (other than the above areas) 2 other raw materials	3/4 (composite textile mills) 1/2 (other mills)	$2^1/_4$	$2^1/_4$
(ii) Man-made fibre	$1^1/_2$	1/2	$1^3/_4$	$1^3/_4$
(iii) Jute textiles	$2^1/_2$	1/4	1 (for domestic sales) and $1^1/_2$ (for exports)	$1^1/_2$
(iv) Rubber products	2	1/4	$1^3/_4$	$1^3/_4$
(v) Fertilizers—				
(a) For nitrogenous plants	3/4 (units near refinery) $1^1/_2$ (units away from refinery)	Negligible	1 (where stocks are in plant site) $1^1/_2$ (where stocks are also in up-country centres)	$1^1/_4$
(b) For phosphatic plants	2(units in port areas) 3(units away from port areas)	Negligible	1 (where stocks are in plant site) $1^1/_2$ (where stocks are also in up-country centres)	$1^3/_4$
(vi) Pharmaceuticals	$2^3/_4$	1/2	2	$1^1/_4$
(vii) Dyes and dyestuffs	$2^1/_4$	1	3/4	$2^1/_4$

EXHIBIT 5.3 (Cont.)

1	2	3	4	5
(viii) Basic industrial chemicals	$2^3/_4$	1/4	1	$1^3/_4$
(ix) Vegetable and hydrogenated oils	1	Negligible	3/4	
(x) Paper	2–6 bamboo and wood (to be built up in stages from November to May and thereafter to be brought down) $2^1/_4$ chemicals	Negligible	1 (for controlled sales) 1/4 (for free sales)	3/4
(xi) Cement	$2^1/_4$ gypsum $1^1/_4$ limestone 3/4 coal $1^1/_2$ packing materials	3/4	1	1
(xii) Engineering automobiles and ancillaries	$2^1/_4$	3/4	$2^1/_2$	$2^1/_2$
(xiii) Engineering consumer durables	2	3/4	$2^1/_2$	$2^1/_2$
(xiv) Engineering ancillaries (other than automobile ancillaries and component suppliers)	2	3/4	$2^1/_2$	$2^1/_2$
(xv) Engineering machinery manufacturers and other capital equipment suppliers (other than heavy engineering)	$2^3/_4$	$1^1/_4$	$3^1/_2$	$3^1/_2$

N.B.: Raw material and stocks in process are expressed as so many months' consumption and cost of production respectively.

* *Financial Express*, 23 August 1975, an extract from Tandon Committee Report, and pp. 20–21 of the report of the Study Group to Frame Guidelines for Follow-up of Bank Credit by Reserve Bank of India.

together in a matrix form so that combining the requirements for different components can be done. Materials planning is best explained by Exhibit 5.4.

EXHIBIT 5.4

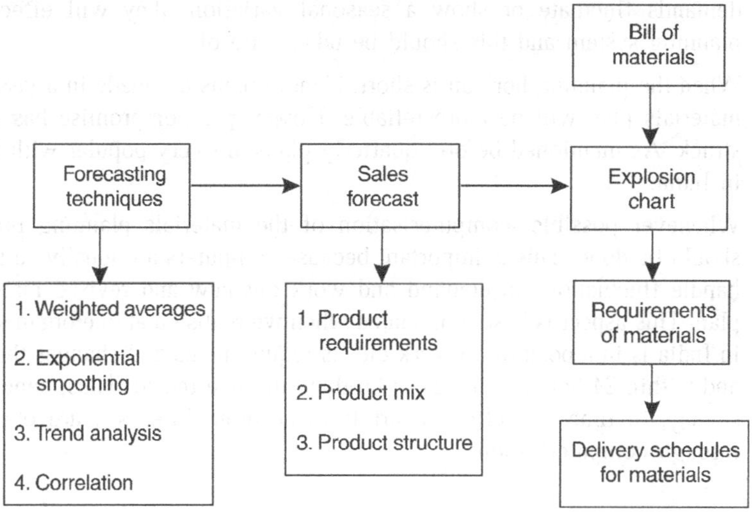

Materials Planning can be used to assess the "firm requirements" for different planning horizons. As the planning horizon exceeds one year, the forecast becomes less reliable. Normally, planning is done on a quarterly basis. This is because the materials planner, at the beginning of every quarter, will find that some materials are in short supply and some in excess, owing to errors in forecasting. Thus, operating on a quarterly basis helps to rectify these errors. At the same time, since quarterly periods are sufficiently long enough in most cases, realistic ordering can be done with the suppliers. This technique is ideally suited for engineering industries.

2. Past Consumption Analysis

For items that are consumed on a continuous basis and for which no bill of materials is possible, this technique is used. Here the past consumption data is analysed and a projection for the future is made, taking into account the past and future production. Certain guidelines are worked out for each item based on the "average" or "mean" consumption and the "standard deviation". These statistical tools, which are very effective for tackling fluctuation in consumption, are explained in the chapter on inventory systems. This technique is particularly used for direct and indirect materials for which no straightforward norms of consumption can be worked out. The technique has been successfully used in process industries.

GUIDELINES

Given below are some of the guidelines which can be used for effective and reliable materials planning.

1. In working out the horizon for planning purposes, the longest lead times that are anticipated must be taken into account so that they are adequately covered.
2. Operating environment of the firm must be carefully analysed. Thus if the demands fluctuate or show a seasonal variation, they will effect the planning system and this should be taken care of.
3. When the planning horizon is short, if more plans are made in a year, the materials plan will be more reliable. However, a compromise has to be struck. As mentioned before, quarterly plans are very popular with firms in India.
4. Whenever possible, computerisation of the materials planning process should be done. This is important because computers are ideally suited to handle fluctuations in demand and work out new and revised materials plan. This aspect is best appreciated when we realise that one organisation in India is in a position to work out its requirements and changes thereof, and within 24 hours a change in product mix is effected. Thus, handsome savings in time and effort, apart from accurate forecasts, are obtained through computerisation.

BUDGETING

Materials budget can be prepared once the requirements are worked out as shown in Exhibit 5.4. The process is represented in Exhibit 5.5.

EXHIBIT 5.5

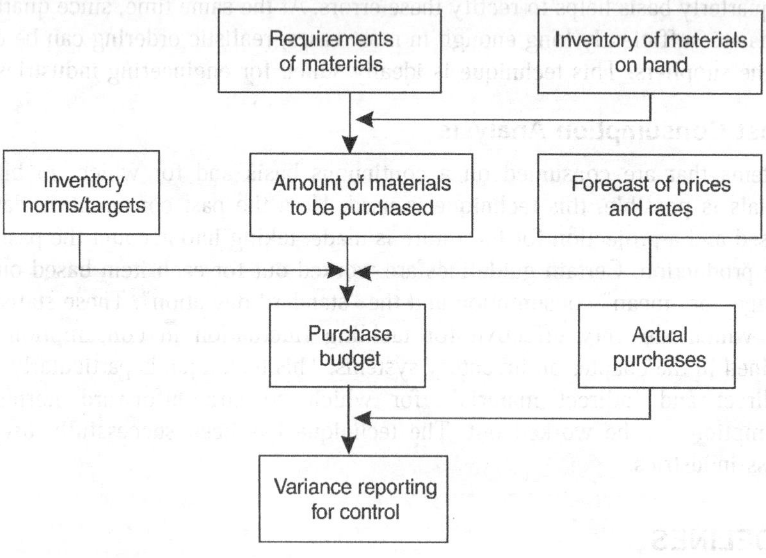

Thus we can easily see that the purchase budget takes into account the inventory on hand and orders on hand. Besides the budget itself may be formulated to attain certain, targeted inventory levels. It is the usual practice to formulate budgets both in terms of quantity and money. The main purposes served by the budget are as follows:

1. The materials management department knows exactly its resource availability so that it can plan its purchases and long term contracts in the most optimum manner taking into account price trends, market position, etc.
2. To record the prices on which the budget was formed and to compare it finally with the actual prices so that controllable and uncontrollable elements are clearly brought out.
3. To help the financial management of materials by clearly stating the cash requirements for the budget period (usually a year) as well as for shorter horizons (monthly or quarterly).

The most important element of budgeting is the review and control aspect. The control aspect should include corrective action with regard to expenditure where the budget target has been exceeded appreciably. The budget must be also adjusted for errors made in its compilation. Periodic statements should be prepared and the budget should be compared with actuals so that major variances can be highlighted for action.

6

ABC Analysis

WHAT IS ABC ANALYSIS?

ABC Analysis is a basic analytical management tool which enables top management to place the effort where the results will be greatest. This technique, popularly known as Always Better Control or the alphabetical approach, has universal applications in many areas of human endeavour. The technique tries to analyse the distribution of any characteristic by money value of importance in order to determine its priority. In materials management, this technique has been applied in areas needing selective control, such as inventory, criticality of items, obsolete stocks, purchasing orders, receipt of materials, inspection, store-keeping, and verification of bills.

The annual consumption analysis of any organisation would indicate that a handful of top high value items—less than 10 per cent of the total number—will account for a substantial portion of about 75 per cent of the total consumption value, and these few vital items are called 'A' items which need careful attention of the materials manager. Similarly, a large number of 'bottom' items—over 70 per cent of the total number called the trivial many—account only for about 10 per cent of the consumption value, and are known as the 'C' class. The items that lie between the top and bottom are called the 'B' category items.

ADVANTAGES OF ABC ANALYSIS

This approach helps the materials manager to exercise selective control and focus his attention only on a few items when he is confronted with lakhs of stores items. By concentrating on 'A' class items, the materials manager is able to control inventories and show 'visible' results in a short span of time. By controlling the 'A' items, and doing a proper inventory analysis, obsolete stocks are automatically pinpointed. Many organisations have claimed that ABC analysis has helped in reducing the clerical costs and resulted in better planning and improved inventory turnover. ABC analysis has to be resorted to because equal attention to 'A', 'B' and 'C' items will not be worthwhile and would be very expensive. Concentrating on all the items is likely to have a diffused effect on all the items, irrespective of the priorities.

MECHANICS OF ABC ANALYSIS

The mechanics of classifying the items into 'A', 'B' and 'C' categories is described in the following steps:

1. Calculate rupee annual issues for each item in inventory by multiplying the unit cost by the number of units issued in a year. It is assumed that the issues and consumption are the same.
2. Sort all items by rupee annual issues *in descending sequence.*
3. Prepare a list from these ranked items showing item no., unit cost, annual units issued and annual rupee value of units issued.
4. Starting at the top of the list, compute a running total, item-by-item issue value and the rupee consumption value.
5. Compute and print for each items the cumulative percentages for the item count and cumulative annual issue value.

The normal items in most organisations show the following pattern:

1. 5 per cent to 10 per cent of the top number of items account for about 70 per cent of the total consumption value. These items are called 'A' items.
2. 15 per cent to 20 per cent of the number of items account for 20 per cent of the total consumption value. These items are called 'B' items.
3. The remaining number of items account for the balance 15 per cent of the total issue value. These items are called 'C' items.

A graph on cumulative value against cumulative per cent of items is prepared (as in Exhibit 6.1) to visually see the ABC categories. Instead of doing ABC

EXHIBIT 6.1

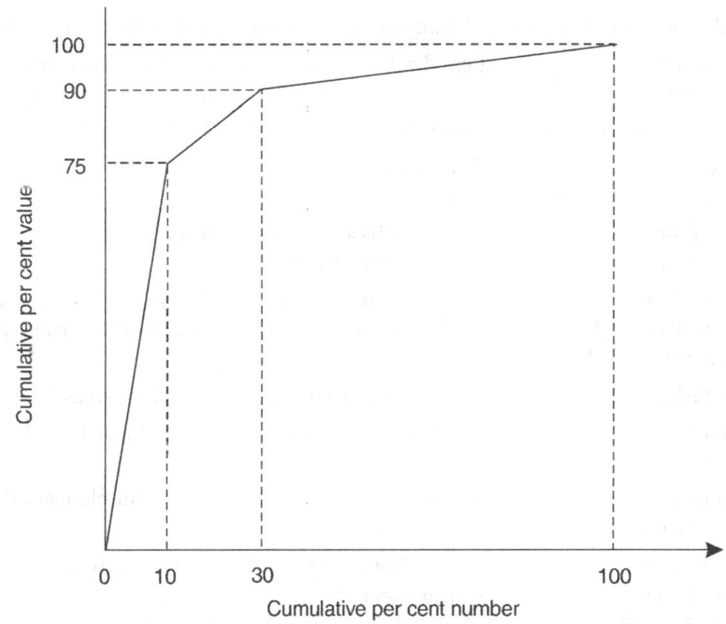

analysis, as discussed earlier, many firms classify items as accounting for consumption value of above Rs. 10,000, 'B' items between Rs 1,000 and Rs 10,000 and 'C' items less than Rs 1,000 of annual consumption. These limits vary depending upon the size of the firm:

The basic principles in ABC analysis are:

(i) The analysis does not depend upon the unit cost of the items but only on its annual consumption value;

(ii) it does not depend on the importance of the item; and

(iii) the limits for ABC categorisation are not uniform but will depend upon the size of the undertaking, its inventory as well as the number of items controlled.

PURPOSE OF ABC ANALYSIS

The object of carrying out ABC analysis is to develop policy guidelines for selective control.

Normally, once ABC analysis has been done, the following broad policy guidelines can be established in respect of each category:

A-items: *High consumption value*	*B-items:* *Moderate value*	*C-items:* *Low consumption value*
1. Very strict control	Moderate control	Loose control
2. No safety stocks (or very low)	Low safety stocks	High safety stocks
3. Frequent ordering or weekly deliveries	Once in three months	Bulk ordering once in 6 months
4. Weekly control statements	Monthly control reports	Quarterly control reports
5. Maximum follow-up and expediting	Periodic follow-up	Follow-up and expediting in exceptional cases
6. Rigorous value analysis	Moderate value analysis	Minimum value analysis
7. As many sources as possible for each item	Two or more reliable sources	Two reliable sources for each item
8. Accurate forecasts in materials planning	Estimates based on past data on present plans	Rough estimates for planning
9. Minimisation of waste, obsolete and surplus (review every 15 days)	Quarterly control over surplus and obsolete items	Annual review over surplus and obsolete material
10. Individual postings	Small group postings	Group postings
11. Central purchasing and storage	Combination purchasing	Decentralised purchasing
12. Maximum efforts to reduce lead time	Moderate	Minimum clerical efforts
13. Must be handled by senior officers	Can be handled by middle management	Can be fully delegated

Colour coding can be used to identify A, B and C categories in stores. Usually red is used for 'A' items, pink for 'B' items and blue for 'C' items.

'A' items merit a tightly controlled inventory system with *constant attention* by the purchase and stores management. A large effort per item on only a few items can cost only moderately, but the effort can result in large savings. 'B' items merit a formalised inventory system with periodic attention by the purchase and stores management. 'C' items use a simpler system designed to cause the *least trouble* for the purchase and stores department, perhaps even at the cost of a little extra inventory cost. It is also common to further subdivide 'A' items as A_1 and A_2 or A^+ and A^- and similarly categories for 'B' and 'C' items exist for exercising finer control.

OBJECTIVE OF ABC ANALYSIS

As discussed earlier, the ABC analysis enables the materials manager to exercise selective control when he is confronted with a large number of items. Tighter and accurate procedures are essential for 'A' value items relating to materials planning, forecasts, ordering, review, records, postings, revisions, lead time analysis, safety stock, materials consumption control, purchase budget, delivery schedule, value analysis, follow-up, clerical efforts, physical stock verification, receipts, issues, stores accounting and inspection. The degree of control should be rigorous for 'A' items and should be minimum for 'C' items.

ABC analysis is also helpful to rationalise the number of orders and reduce the overall inventory. For instance, let us consider three items accounting for a consumption of Rs 60,000, Rs. 4,000 and Rs 1,000 respectively, each being ordered four times a year. The following pattern would then emerge:

EXHIBIT 6.2

Item no.	*Annual consumption value (Rs)*	*Number of orders*	*Value per order*	*Average inventory*
1	60,000	4	15,000	7,500
2	4,000	4	1,000	500
3	1,000	4	250	125
	Total inventory			Rs 8,125

If A, B, C analysis has been done, then it implies that 'A' item should be ordered more, say eight orders, three for 'B' and one for 'C' item. Then the pattern given in Exhibit 6.3 emerges.

EXHIBIT 6.3

Item no.	*Annual consumption value (Rs)*	*Number of orders*	*Value per order*	*Average inventory*
1	60,000	8	7,500	3,750
2	4,000	3	1,333	667
3	1,000	1	1,000	500
	Total inventory			Rs 4,917

It is thus noted that even though the total number of orders is the same, the average inventory has been reduced significantly by recognising that 'A' items should be ordered more frequently. (The optimum number of orders can be arrived at by formulae discussed under Inventory Systems.)

The details of ABC classification relating to an engineering factory are given in Exhibit 6.4. High value consumption items have been segregated as A_1 and A_2 for purposes of better control. Similarly, the low consumption value 'C' items have been grouped into C_1 and C_2 classes, because of the large number of items involved. The classification has been done on the overall basis as well as on major groups for purposes of control. The inventory position for each sub-classification has also been indicated in the table in order to compare the consumption and inventory value. For instance, in the engineering factory mentioned above, 8 per cent of the number of items constituted the 'A' category accounting for 68 per cent of the value, 17 per cent of items 'B' category approximating to 20 per cent of the value and the remaining items forming the 'C' category constitute 12 per cent of the value.

LIMITATIONS OF ABC ANALYSIS

ABC analysis, in order to be fully effective, should be carried out with standardisation and codification. ABC analysis is based on grading the items according to the importance of performance of an item, that is by V.E.D.—vital, essential and desirable—analysis discussed later. Some items, though negligible in monetary value, may be vital for running the plant, and constant attention is needed. If the inventory position is analysed according to the value, commonly known as XYZ analysis, then results of ABC and XYZ analysis will be different, depending upon the nature of obsolete items. The results of ABC analysis have to be reviewed periodically and updated. It is a common experience that a 'C' item, like diesel oil in a firm, will become the most high value item during power crisis. However, ABC analysis is a powerful approach in the direction of cost reduction as it helps to control items with a selective approach.

EXHIBIT 6.4
ABC ANALYSIS BASED ON ANNUAL CONSUMPTION (Value Rs lakhs)

Stock code	Description	*A items (above Rs 10,000)*																	
		A_1 items (above Rs 50,000)										*C items (up to Rs 1,000)*							
		Above Rs 1,00,000		*Rs 50,000– Rs 10,000*		*Total A_1*		*A_2 items Rs 10,000–50,000*		*Total A ($A_1 + A_2$)*		*B items Rs 1,000–10,000*		*C_1 items Rs 250–1,000*		*C_2 items up to Rs 250*		*Total C ($C_1 + C_2$)*	
		No.	*Value*	*No.*	*Value*	*No.*	*Value*	*No.*	*Value*	*No.*	*Value*	*No.*	*Value*	*No.*	*Value*	*No.*	*Value*	*No.*	*Value*

Stock code	Description	Above Rs 1,00,000 No.	Value	Rs 50,000–Rs 10,000 No.	Value	Total A_1 No.	Value	A_2 No.	Value	Total A No.	Value	B No.	Value	C_1 No.	Value	C_2 No.	Value	Total C No.	Value	Total items No.	Value
0	Raw materials	31	82.42	41	28.18	72	110.60	106	24.90	178	135.50	227	8.61	162	0.87	285	0.23	447	1.10	852	145.21
1	Shop stores	2	4.55	4	3.20	6	7.75	24	4.96	30	12.71	70	2.43	90	0.48	198	0.16	288	0.64	388	15.78
2	Standard parts	–	–	–	–	–	–	10	1.89	10	1.89	95	2.59	193	0.96	905	0.52	1,098	1.481	203	5.96
3	Cutting tools	–	–	–	–	–	–	3	0.39	3	0.39	88	2.09	146	0.74	354	0.29	501	1.03	592	3.51
4	Measuring tools	–	–	–	–	–	–	–	–	–	–	8	0.13	28	0.14	32	0.03	60	0.17	68	0.30
5	Sundry tools	–	–	–	–	–	–	–	–	–	–	17	0.46	18	0.10	144	0.06	162	0.16	179	0.62
6	Equipment spares	–	–	–	–	–	–	4	0.57	4	0.57	57	1.55	127	0.69	588	0.33	685	1.02	746	3.14
7	Electrical stores	–	–	–	–	–	–	8	1.00	8	1.00	57	1.41	97	0.48	616	0.27	713	0.75	778	3.16
	Sub-total	33	86.97	45	31.38	78	118.35	155	33.71	233	152.06	619	19.27	862	4.46	3,092	1.89	3,854	6.35	4,806	177.68
	Percentages	0.7	48.8	0.9	17.61	6	66.4	3.2	18.8	4.8	85.2		14.7	10.818	2.5	64.0	1.0	82	3.5	100	100
8	Bought-out	14	31.44	10	7.26	24	38.70	111	23.57	135	62.27	353	12.65	122	1.25	234	0.22	456	1.47	944	76.29
9	Bought-out components	5	6.46	5	3.09	10	9.55	53	11.18	63	20.73	180	6.74	111	0.61	227	0.19	338	0.80	581	28.27
10	Sub-total	19	37.90	15	10.35	34	48.25	164	34.75	198	83.00	533	19.39	233	1.86	461	0.41	794	2.27	1,525	104.56
	Grand total	52	124.87	60	41.73	112	166.60	319	68.46	431	235.06	1,152	38.66	1,195	6.32	3,553	3.20	4,748	8.62	6,331	282.34
	Percentages	0.8	44.3	0.9	14.8	1.7	59.1	4.9	24.6	6.6	83.7	18.2	13.7	18.8	2.2	56.1	0.8	75	3.0	100	100

7

Codification and Standardisation

NATURE OF CODIFICATION

Any organisation stores a large number of items, often running into lakhs. There should, therefore, be some means of identifying them. A common practice is to describe them by individual names. Since several departments use the same item, they call the same item by different names and store them in different places. One store-keeper might classify an item as Sal Ammoniac, whereas a research chemist might identify it under the name of Ammonium Chloride, only to be told that it is not available. A classic example comes from the U.K.: An electrical firm found that a simple item of a screw with a width of 3/8" and length of 6" had as many as 118 names depending on the type of usage and the department using the screw. A few names were plunger, dowel pin, roller, locating peg, drive pin, pinion spindle, pin mould holding, motor drive spindle, trip arm pin, armature stud, etc. (see Exhibit 7.1). A proper codification removed the 117 stocking points and with it the associated inventories, entries, obsolete item, etc.

EXHIBIT 7.1
118 NAMES DESCRIBING PLAIN SINGLE DIAMETER PINS

Pin
Pin hinge
Peg
Locating peg
Dowel
Valve cotter
Canopy rod
Roller shaft
Pivot
Brush
Plug
Spindle
Stop
Roller
Plunger

Carrier link pin
Standing valve
Valve cotter pin
Clutch toggle pin
Swing bolt pin
Pin-towing eye
Wheel stud dowl
Inlet rocker bracket
Oilway plug
Armature cradle bearing rod
Connector moving oscillator coil
Main spindle location pin
Latch
Button
Stud

EXHIBIT 7.1 (Cont.)

Connector	Rod
Trunnion	Hinge rod
Electrode	Lock rod
Pin pivot	Push rod
Pin anchor	Valve rod
Pin crank	Plain rod
Drive pin	Rivet
Pivot top	Spacer
Plain pin	Post
Dowel pin	Bar
Wire pin	Dashpot rod
Bond pin	Valve stem
Sealing sleeve	Flag pivot
Driving pin	Crank rivet
Intercooler fitter dowel	Roller pin
Starting handle	Pin link
Shaft pin	Post mileage
Distributor drive	Dashpot rod
Plain pin	Valve stem
Water pump driving pin	Flag pivot
Governor drive	Crank rivet
Coupling pin	Roller pin
Manhandling bar knuckle	Pin link
Jaw pin	Post mileage
Pump drive	Location pin
Centre driving peg	Operating pin
Slave spindle	Trip arm pin
Clutch roller	Pawl spindle
Counter shaft	Special pin
Reset spindle	Eyebolt pin
Drain tap boss	Bearing pin
Piston rod pin	Driving bar
Pin push lever	Lever spindle
Fuel pump tuppet	Armature stud
Kicket catch pin	Pinion spindle
Thrust collar pin	Water pump dowel
Jack leg plug	Chain anchor dowel
Dowel form main bearing	Oil relief ball stop
Reset frame shop pin	Contact lever axle
Breather hole plug	Pin clamp lever axle
Selector bar plug	Fleat lever stud
Governor weight spindle	Pin mould holding
Connecting link rod	Motor drive spindle
Roller spindle	Sanding valve roller

PROCESS OF CODIFICATION

Codification is a process of representing each item by a number, the digits of which indicate the group, the subgroup, the type and the dimension of the item. Many organisations in the public and private sectors, railways and DGS & D, have their own system of codification, varying from eight to thirteen digits. The first two digits normally represent the major groups, such as raw materials, spare parts, sub-contracted items, hardware items, packing material, tools, oil, stationery, etc. The next two digits indicate the sub-groups, such as ferrous, non-ferrous, etc. Dimensional characteristics of length, width, head diameter usually constitute the further three digits and the last digit is reserved for minor variations.

The codification process could be obtained by the nature of items in grouping all items of the same metal content, say ferrous, non-ferrous, etc. The system of codification could be built by the end use of items, that is, items grouped according to maintenance, spinning, weaving, packing, foundry, machine shop, etc. The codification could be thought of on the basis of source of purchasing where items obtained from one source of purchasing are grouped together and given codes. The codification could also be built on the basis of alphabetical listing.

It can also be done by numerical digits or by a combination of English letters and numericals. The universal decimal classification of codification, which is used in libraries, introduces decimals for identifying items.

Whatever may be the basis, each code should uniquely represent one item. It should be simple and capable of being understood by all. Codification should be compact, concise, consistent and flexible enough to accommodate new items. As far as possible uniform dimensions, say the metric system, should be adopted. It should be meaningful and oriented towards the needs of an individual organisation. The groupings should be logical, holding similar parts near to one another. Each digit must be significant enough to represent some characteristic or aspect of the item. Two fundamental systems are described in detail in the following paragraphs.

KODAK SYSTEM

The Kodak system has been developed by Eastman Kodak Company of New York which consists of 10 digits of numerical code. The logic of major grouping is based on sources of supply. All materials are divided into 100 basic classifications, contributed only by procurement considerations. For instance, a bolt is listed as hardware item if this is listed in hardware catalogues and available with hardware suppliers. If this bolt, however, is available only as part of the machine, it will be available under maintenance. Each class is divided into 10 sub-classes. For example, if class 20 represents cutting tools then 200 represents drills, reamers, counter bars, etc.

BRISCH SYSTEM

The Brisch system named after a prominent consulting engineer in the UK consists of seven digits and is applied in three phases. The items are grouped into suitable preliminary categories, such as assemblies, sub-assemblies, components, off the shelf items. After these preliminary categories, items are grouped within the respective clans in order to bring similar items together. The Brisch system, though it consists only of seven digits, is quite comprehensive as the basis is on logical major groupings.

ADVANTAGES OF CODIFICATION

As a result of rationalised codification, many firms have reduced the number of items. It enables systematic grouping of similar items and avoids confusion caused by long description of the items. Since standardisation of names is achieved through codification, it serves as the starting point of simplification and standardisation. It helps in avoiding duplication of items and results in the minimisation of the number of items, leading to accurate records. Codification enables easy recognition of an item in stores, thereby reducing clerical efforts to the minimum. If items are coded according to the sources, it is possible to bulk the items while ordering. To maximise the aforesaid advantages, it is necessary to develop the codes with all concerned, namely, personnel from design, production, engineering, inspection, maintenance and materials.

NEED FOR STANDARDISATION

Standardisation is required not only for ensuring procurement of the right quality of incoming material, but also for cost reduction. It will be remembered that considerable publicity was given to the large-scale cost reduction in materials cost through standardisation by the US defence services under the guidance of the then Defence Secretary, Robert S. McNamara.

STANDARDISATION IN INDIA

Effective steps have been taken in a number of organisations in India for cost reduction through reduction in the number of stores items. The aim of standardisation should be to have uniform standards for similar items, and the standards evolved should take cognizance of the indigenous availability of materials to the maximum extent possible. Realising this importance, the Indian Standards Institution (ISI) has promoted over 7,000 standards covering raw materials, components and finished products. These are widely publicised and are readily available from the Indian Standards Institution.

IMPORTANCE OF STANDARDISATION

Standardisation enables the materials manager to achieve overall economy and ensures inter-changeability of parts. Since more than one manufacturer can supply standard items, it will imply better availability, better price and better delivery. Standardisation also implies routinising purchase efforts, less stock and hence less obsolete items. It also means less inspection efforts; as a matter of fact, many organisations do not check routine items bearing the ISI mark in a detailed manner and resort to inspection of only a small fraction of items. It is also possible to enter into rate/running contract with standard items. At the suppliers' end, manufacturing into economic lot sizes is not a problem for standard items.

While dealing with Variety Reduction and Standardisation, the Committee on Public Undertakings* mentions the following:

"Another reason for large inventories in public undertakings was that the items of stores were of multitudinous variety. Thus in Oil and Natural Gas Commission it was found that in the same project different makes of machines were being used for identical purposes. This was true not only of imported machinery and equipment but also of indigenous makes. For instance, in the case of 3 ton trucks, Bedfords, Fargos and TMBs were all in use, simultaneously in most of the projects. The result was that each major project had to stock about 3,000 to 4,000 categories of spare parts pertaining to motor vehicles alone.

"Similarly in Hindustan Insecticides it was noted that at the end of 1964–65 there were too many varieties for certain items of stores, e.g., Valves—150 varieties, Ball and Roller Bearings—64 varieties, MS bolts and nuts—44 varieties, Drills H.S. (S.S. & T.S.)—36 varieties etc.

"In yet another undertaking, namely, Fertilizers & Chemicals Travancore Ltd., there were nearly 100 varieties of electric motors. There were too many varieties of tools, oils, paints, enamelled copper-wire electrodes, heat exchangers, pumps, oil seals, valves, ball-bearings, various hardware items, printing and stationery items and so on. The committee therefore desires that the matter should be pursued vigorously by public sector undertakings in coordination with Indian agencies."*

DEFINITION

A standard is defined as a model or general agreement of a rule established by authority, consensus, or custom, created and used by various levels of interest. For instance, an individual may be the starting point of using the standard and then his department will use the same standard to suit its needs. The firm may similarly prepare, by consulting different departments, a standard for guiding the activities. Related industries in the industry group may also prepare industrial standards. At the national level, by consulting manufacturers, scientists, users and government departments, national standards are evolved. Such national standards are presented and discussed to form international standards. The standards could cover a variety of industries, such as engineering, textile, chemical, pharmaceutical, agricultural as also education. The topics covered under standards can include purchase contracts, forms, sampling, testing, safety measures, etc. Thus, standardisation can be conceived on a three-dimensional plane of levels, industries and topics.

---

* *Fortieth Report on Materials Management in Public Undertakings: Committee on Public Undertakings*, page 14, Third Lok Sabha, Lok Sabha Secretariat, New Delhi.

SIMPLIFICATION

The process of standardisation logically leads to simplification or variety reduction. This implies reducing unnecessary varieties and standardising to the most economical sizes, grades, shapes, colours, types of parts and so on. In large organisations handling 50,000 items, there are several items under stock having very little variation in quality, dimension or functional effectiveness. Nevertheless these continue to be in stock for historical reasons. Practical experience indicates that often this plethora of variety is not even perceptible.

These items can be analysed by their frequency of usage or movement analysis over the last few years. Frequency or movement analysis would bring out items which are seldom used or not used at all. On the basis of this analysis, a company could set standards to replace these items. The setting up of a standard depends on the effect the dimension variation has on the performance of the product. To illustrate let us take a look at the frequency figures below. It is evident that for the same variation from the optimum size B, the performance of the product in Exhibit 7.2 is less affected than in Exhibit 7.3. Therefore, for items which affect the product performance as in Exhibit 7.2 a great deal of flexibility is introduced in fixing up standards with a firmer quality control where the standards are near A or C for an item.

EXHIBIT 7.2

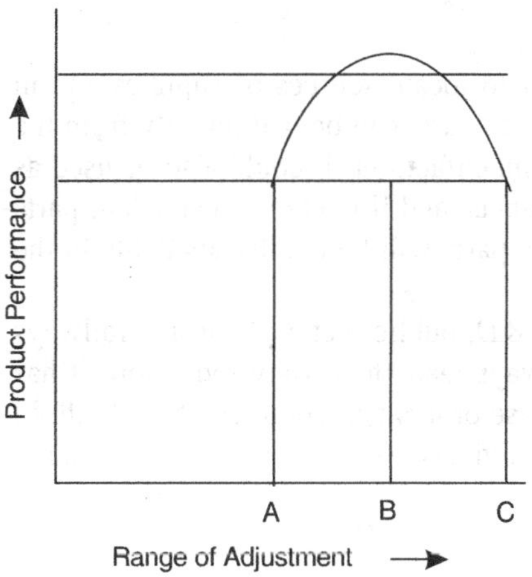

EXHIBIT 7.3

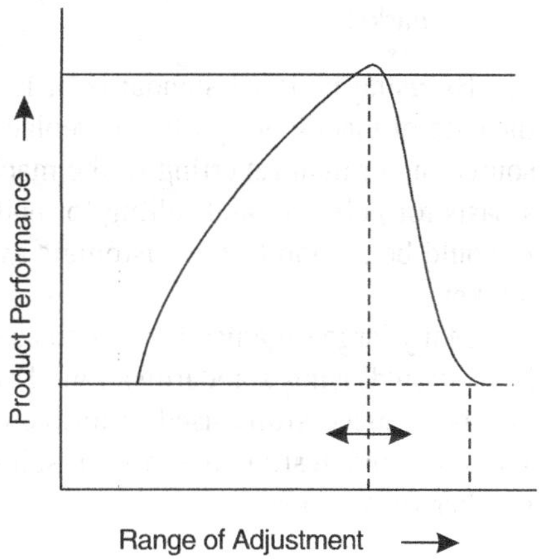

PREFERRED NUMBERS

We are all aware that bulbs of 15, 25, 40, 60, 100 and 150 watts are available. It is interesting to note that these numbers have been worked out on the basis of what is known as an R_5 series. They are known as "Preferred numbers". This series is a geometric progression with its starting number as $5\sqrt{10}$ = 15 and its fifth number 150 is 10 times the first number which is 15. Similarly, one could have a R_{10}, R_{20}, . . ., R_{80} series. These are series which increase 10, 20, . . ., 80 times in the 10th, 20th, . . ., 80th steps, respectively, the increments being based on $10\sqrt{10}$, $20\sqrt{10}$, . . ., $80\sqrt{10}$. The choice of a series depends on the range of dimensions to be covered. Similar logic is applicable in the case of currency, such as rupee notes of denomination 1, 2, 5, 10, 20, 100, 1000, 5000 and 10000. It is not essential to use all preferred numbers in a series. Some could be omitted.

Preferred numbers can be pre-determined by manufacturers in advance of production, whereas frequency analysis is based on data over a fairly long period. However, preferred numbers cannot be used where dimensions are not involved. A good example would be colours in the textile industry.

BENEFITS OF STANDARDISATION

The important benefits are as follows:

1. Standardisation helps reduce inventory items.
2. It helps in evolving better means of communication about an item in the company.
3. It forms a base for further inventory analysis.
4. The specification of items can be more clearly spelled out, making quality control firm.
5. In a developing economy like ours, where the need is to promote exports, insistence on standards helps in creating confidence in the international market.

By using national standards, it is easier to locate sources of supplies and in the case of machinery parts, the replacement of parts can be had directly from the source rather than reverting to the machine manufacturer. It could also be used as a basis for publicity and selling for end-products, and if made from standard parts it would be a boon to the customers as these parts will be readily available in the market.

Many large organisations, such as DGS &D, public sector plants and railways have started using standardisation. This always leads to variety reduction. It has also been successfully used in forms, purchase orders and contracts. Standardised sampling and testing has also resulted in reduced inspection costs in a large number of organisations.

8

Source Selection

WHY SOURCE SELECTION?

Reputed suppliers are intangible assets to any organisation. For they are not only suppliers of materials but are also extremely important sources of information with regard to market conditions, price trends and the general industrial climate. It is, therefore, natural that many organisations have accepted source selection as a corporate policy. This helps in bringing about a fair competition among the suppliers and supply failures are kept at a minimum. Source development is also important for import substitution, cost reduction and quality improvement. Source development needs are dependent on factors such as make or buy decisions, amount of sub-contracting, break-even points at manufacturing and plant capacity. It should, however, be remembered that source selection, development and presentation is a continuing activity.

STAGES IN SOURCE SELECTION

Broadly, the gamut of source selection activities can be divided into the following:

Searching

The need for a material or product is the starting point. Some firms use forms such as the one in Exhibit 8.1. The search process begins with the finalisation of specifications in consultation with technical departments. It consists in identifying the sources of supply—the initial survey should be exhaustive. Normally, comprehensive information can be obtained from the *Indian Trade Journal*, newspaper advertisements, telephone directory as well as from the publications of Government agencies like the Directorate General of Technical Development, Directorate General of Supply and Disposal and the Indian Standards Institution. The routine sales-calls by suppliers' representatives is another important source of information. Salesmen should be encouraged to present the products in which the organisation is presently interested, and their company's entire product line. This enables the buyer to identify potential sources for products which under normal circumstances might be overlooked. Trade-shows, technical exhibitions and samples are other important source that assist the search process.

EXHIBIT 8.1

To ____________________

From ____________________ Date ______________

REQUEST FOR SOURCE DEVELOPMENT

Reason for Request:	New item/New product/Buy not make/ Price/Quality/Delivery/Import Substitution/ Only one source/Other
Component Code No:	______________
Component Drawing No:	______________ (Please attach 5 copies)
Present Annual Requirement:	Qty: ________ Value: ________
Present Price Paid:	Rs
Present Vendors with Address:	State clearly nature of problem with vendor:
a. ____________________	____________________
b. ____________________	____________________
____________________	____________________
c. ____________________	____________________
____________________	____________________
Used on which Product?	____________________
When Required for Assembly/Production:	Week ________ Date ________
Number of Samples Attached:	Local ________ Imported ________
What Assistance can we give Vendor?	Tooling/Testing/Materials composition/ Method of manufacture/Other
Further Technical and other Details:	____________________

Selection

The process of search thus provides a list of all possible sources, which has to be trimmed down. The focus here is on more specific information on the supplier's financial strength, quality, facilities, efficiency, industrial relations, technical excellence and position in industry. In other words, we want to identify the possible sources and select such sources which will meet the delivery and quality requirements in the long run at competitive price levels. The selection of suppliers starts with the floating of enquiries by the buyer to the possible sources that are made available through the search process. Certain progressive suppliers often contact buyers with a request to be included in the buyers' list of approved suppliers. Both the buyers' enquiry and the vendors' request to be considered as a regular supplier, normally lead to visits by the vendors' sales representative. The buyer gets an opportunity to present his needs in greater details and also to assess vendors' personnel and capabilities in meeting his needs. The buyer follows this by meeting the technical personnel of the vendor, and inspecting the vendor's plant to assess the technical capabilities, efficiency, equipment, financial viability, quality control, raw material practices, and general management aspects.

Normally the buyer consolidates such information in a systematic manner as shown in Exhibit 8.2. The buyer may also write in confidence to some of the organisations which may be using the materials supplied by the vendor in order to assess vendor's rating with his clients. The information thus collected and culled provides the basis for a preliminary selection of suppliers with whom trial orders can be placed after negotiations.

Negotiations and Trial Orders

After the preliminary screening and selection, starts the process of negotiation with the vendors for placing trial orders. Correct and cordial relations with the vendors are essential for mutual cooperation. The vendors should be provided with the same courtesies that an organisation expects for its own salesmen or gives to its customers. Various aspects, including terms of delivery, price and quality are finalised during negotiations and then the purchase orders are released for the initial trial. Normally, the trial orders do not exceed more than 20 to 30 days' requirements as the buyer has to be convinced, in the first place, about the vendor's capability in meeting his needs.

There are many factors which enhance the process of negotiation. They include:

1. The need should be described fully but concisely—quantity requirements and other technical details.
2. Personal contacts between the executives of the buyer and the vendor often help in maintaining a smooth relationship and understanding of each other's problem-areas.
3. On a regular basis, the buyer may organise a conference of his suppliers with a view to giving them an opportunity to find out where their materials are utilised, to compare and clarify the quality and other technical requirements of the materials they supply.

Rating

After the trial orders are executed, it becomes necessary for the buyer to rate the vendors to enable him to determine how he should apportion his requirements among the vendors. In other words, the basic responsibility of the materials

EXHIBIT 8.2
VENDOR EVALUATION FOR APPROVED LIST

1. Building and General Impression of Good Maintenance	
(a) Check floor space available against that stated in questionnaire form	Same/Different
(b) What material handling equipment is used?	Crane/Hoist/Fork-lift/ Other
(c) Level of housekeeping, cleanliness and maintenance order	Poor/Good/Very good
(d) Age of the plant	5 to 10/15/more old/ modernised/new and modernised
(e) Is care and maintenance of plant a planned programme?	Yes/No
(f) Does the presence of clever methods and set-ups show adaptability in the type of work being done?	Yes/No
(g) Is there space for expansion?	Yes/No
2. Production and Technical Competence	
(a) Check list of machines attached to the questionnaire form	Same/Diff. (Why?)
(b) No. of technically qualified engineers/staff and level of qualification	E.E./M.E./I.E./Other
(c) No. of skilled workers and type of training, i.e., trade apprenticeship, etc.	
(d) Can design assistance be given by the vendor?	Yes/No
(e) Is any basic research being done?	Yes/No
(f) What other work is in progress?	
(g) General level of precision and finish of work in progress compared to what his machines can achieve	Poor/Good/Very good
(h) Is any production planning method used?	Yes/No
(i) Are job orders followed up?	Yes/No
(j) Are there regular contracts for raw material?	Yes/No
(k) Limitations of work due to the size or capacity of the tools and machine?	Yes/No
3. Quality Control and Inspection Facilities	
(a) Is there a regular quality control dept/section?	Yes/No
(b) Are gauges and instruments used for quality control and inspection?	Yes/No
(c) Are gauges and instruments calibrated and corrected on a regular cyclical plan?	Yes/No

(d) Are gauges bought or borrowed?	Bought/Borrowed
(e) Are SQC methods used? If so what is the sampling plan?	Yes/No
(f) Are records kept of the results of instrument checks and production inspection?	Yes/No
(g) Is incoming materials inspected and evaluated?	Yes/No
(h) To whom does the head of QC report?	
(i) No. of people full-time on inspection	Trained/Untrained
(j) What is present per cent rejection?	(5/5–10) 10 years
4. Financial Resources	
(a) What is the approximate value of work in progress?	Rs
(b) What is the approximate value of raw materials and finished goods?	Rs
(c) Has vendor any raw material licence or allocation? If so how much of which material(s) and from which source?	Yes/No
(d) What is the approximate annual turnover of sales?	Rs
(e) What were last year's approximate profits (gross)?	%/Rs
(f) Does he have any future plans for expansion/ diversification?	Yes/No
(g) Names of associate companies	Yes/No
5. Other Companies Supplied to	
(a) Check names mentioned in the questionnaire form	Same/Different
(b) What work was done for these companies?	
(c) How long has vendor been supplying to these companies?	
(d) Does vendor do entire job or does he further subcontract?	Entire job/Further sub-contract
(e) If so, list names of parties work sub-contracted to:	
(f) Timing performance of delivery of a few typical cases	Poor/Good/Very good
(g) Quantity performance of delivery of a few typical cases	Poor/Good/Very good
(h) Obtain reference from other companies if large order is to be given	
6. General Observations	
(a) How would you rate the morale of the total organisation?	Poor/High/Very high
(b) How would you rate the progressiveness and competence of vendor's management?	Poor/Good/Very good
7. Conclusion	
Should this vendor be added to our list of approved suppliers?	Yes/No

manager is not only to locate the sources but also to preserve them through continuous rating. A vendor's performance in meeting the quality, delivery and price standards set by the buyer has to be assessed in a systematic manner. Several rating systems are available. Some aspects can be objectively rated whereas some cannot, but they are nevertheless to be considered while evaluating the vendors. Two vendor rating systems are shown in Exhibits 8.2 and 8.3. Computed on a periodic basis, vendor rating provides the basis for comparing one vendor against the other for the purpose of eliminating the vendors who repeatedly fail to meet the standards. The vendor's representative, in some cases, can be called for a frank discussion to point out the areas of improvement, so that he may continue his association with the buyer's organisation. Such a constructive approach, based on a rating system, will definitely help in improving the performance of the supplier.

EXHIBIT 8.3
VENDOR PERFORMANCE RATING

Appendix 'A'

Name of vendor: P.O. No.

Date

Item

Value Cat.

---

1. Delivery Performance:

(a) *Adherence to time schedule* (Weightage 40%)

$$\frac{\text{No. of deliveries made on the scheduled date}^{*} \times 40}{\text{Total number of scheduled deliveries}}$$

(b) *Adherence to quantity schedule* (Weightage 20%)

$$\frac{\text{No. of correct lot sizes delivered}^{\dagger} \times 20}{\text{Total number of scheduled deliveries}}$$

2. Quality Performance (Weightage 40%):

$$\frac{\text{Quantity accepted} \times 40}{\text{Total quantity supplied}^{\ddagger}}$$

Total

Assessed by Approved by

---

* Deviations with ± 7 days to be ignored.

† Deviations with ± 10 per cent to be ignored.

‡ Rejected quantity resupplied after rectification/sorting out to be treated as fresh supply.

SPECIAL ASPECTS IN SOURCE SELECTION

Local Sources

Generally local sources are preferred, provided they meet the requirements of the buyer in terms of quality, delivery and price. A personal follow-up or a local "phone call" is enough for effecting quicker deliveries. In fact, some organisation prefer local suppliers even though they may not compare favourably with outside suppliers. This local goodwill is sure to pay in the long run. It is not uncommon to find big organisations in India encouraging ancillary suppliers in and around their plants through technical and financial assistance.

Subcontracting

Selection of sources for subcontracting is another special issue. The manufacturer may decide to 'off-load' some products to suppliers so as to facilitate scheduling of production in the plant. It also enables better utilisation of manpower. This is because subcontracting is just an extension of the plant's activities and since subcontractors are generally small, they may be in need of technical assistance. The various advantages and disadvantages of subcontracting have already been dealt with under Make or Buy decision.

Small or Big?

Often in selecting suppliers, the materials manager is faced with this question. If the requirements are small, the small supplier is the obvious choice. A personalised contact can be established and because he is small, he is likely to have fewer customers, will be more loyal and will meet even unusual requests. Big suppliers are equipped to meet larger requirements. They usually have reserve facilities and may thus be able to meet additional requirements. They may have R & D facilities which will enable the buyer to obtain technically superior products.

Part II
PURCHASING MANAGEMENT

9

Creative Purchasing

IMPORTANCE

Purchasing has come to stay as the most important function of materials management. The moment a buyer places an order he commits a substantial portion of the finance of the corporation which affects the working capital and cash flow position. He is a highly responsible person who meets various salesmen and thus can be considered to have been contributing to the public relations efforts of the company. Thus, the buyer can make or mar the company's image by his excellent or poor relations with the vendors.

GOALS OF PURCHASING

The basic objective of the purchasing function is to ensure continuity of supply of raw materials, sub-contracted items and spare parts and at the same time reduce the ultimate cost of the finished goods. In other words, the objective is not so much to procure the raw materials at the lowest price but to reduce the cost of the final product. For ensuring this, there are a large number of well known parameters such as right price, right quality, right contractual terms, right time, right source, right material, right place, right mode of transportation, right quantity and right attitude. All these have to be considered jointly. A diagram indicating these is shown in Exhibit 9.1.

1. Right price

It is the primary concern of any manufacturing organisation to get an item at the right price. But right price need to be the lowest price. In this context it may be worth remembering John Ruskin's famous statement: "There is hardly anything in the world that somebody cannot make a little cheaper and the man who considers price alone is the lawful prey". While it is very difficult to determine the right price, general guidance can be had from the cost structure of the product. The 'tender system' of buying is normally used in public sector organisations but the objective should be to identify the lowest "responsible" bidder and not the lowest bidder. The technique of 'learning curve' also helps the purchase agent to determine the price of items with high labour content. The price can be kept low by proper planning and not by 'rush' buying. Price negotiation also helps to determine the right prices.

EXHIBIT 9.1
PURCHASE PARAMETERS

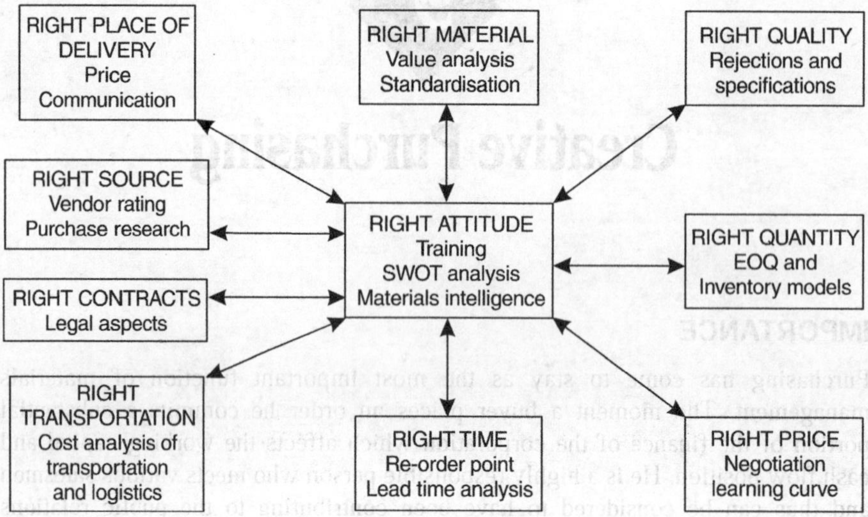

2. Right Quality

Right quality implies that quality should be available, measurable and understandable as far as practicable. In order to determine the quality of a product, sampling schemes, as discussed in the chapter on incoming materials inspection, will be useful. The quality particulars are normally obtained from the indents, and experience indicates that a substantial portion of the indents prepared by the user departments are invariably incomplete. Such incomplete indents often cause unnecessary delays in procurement as the indentor has to be referred to, and if not referred results in heavy rejection. Drawings are also attached to the indents, particularly for spare parts. Since the objective of purchasing is to ensure continuity of supply to the user departments, the time at which the material is provided to the user department assumes great importance.

3. Right Time

For determining the right time, the purchase manager should have lead time information for all products and analyse its components for reducing the same. Lead time is the total time elapsed between the recognition of the need of an item till the item arrives and is provided for use. Obviously, this covers the entire duration of the materials cycle and consists of pre-contractual administrative lead time, manufacturing and transporting lead time, and inspection lead time. Since the inventory increases with higher lead time, it is desirable to analyse each component of the lead time so as to reduce the first and third components which are controllable. While determining the purchases, the buyer has to consider emergency situations like floods, strikes, etc. He should have "contingency plans" when *force majeure* clauses become operative, for instance, the material is not available due to strike, lock-out, floods, and earthquakes. However, rush purchase should be resorted to only in exceptional cases.

Right Source

The source from which the material is procured should be dependable and capable of supplying items of uniform quality. The buyer has to decide which item should be directly obtained from the manufacturer. Aspects such as source selection, source development and vendor rating are discussed in detail in the next chapter on buyer-seller relationships. In emergencies, open market purchases and bazar purchases are resorted to.

Techniques such as value analysis will enable the buyer to locate the right material. Right modes of transportation have to be identified as this forms a critical segment in the cost profile of an item. It is an established fact that the cost of the shipping of ore, gravel, sand, etc., is normally more than the cost of the item itself. Specifying the right place of delivery, say, head office or works, would often minimise the handling and transportation costs. Similarly, packaging forms an important aspect in the cost of an item; for instance, in tooth paste the tube is costlier than the paste it holds.

Right Quantity

The right quantity is the most important parameter in buying. Concepts, such as economic order quantity, economic purchase quantity, fixed period and fixed quantity systems, will serve as broad guidelines. But the buyer has to use his knowledge, experience and common sense to determine the quantity after considering factors such as price structure, discounts, availability of the item, favourable reciprocal relations, and make or buy consideration.

Developing the right attitude, too, is necessary as one often comes across such statements: "Purchasing knows the price of everything and value of nothing"; "We buy price and not cost"; "When will our order placers become purchase managers?"; "Purchasing acts like a post box". Purchasing should, therefore, keep 'progress' as its key activity and should be future-oriented. The purchase manager should not follow the safe and well-trodden path; he should be innovative and his long-term objective should be to minimise the cost of the ultimate product. He will be able to achieve this if he arms himself with techniques such as value analysis, materials intelligence, purchase research, SWOT analysis, purchase budget, lead time analysis, etc.

The buyer has to adopt separate policies and procedures for capital and consumer items. He should be able to distinguish between indigenous and international purchasing procedures. He should be aware of the legal and contractual aspects in international practices.

NEGOTIATION

Negotiation, particularly for prices, has been accepted as a policy by some materials managers, since perfect competition does not exist in Indian situations and hence the concept of lowest bid does not always work. Negotiation is, indeed, a battle of wits and an art of embodying sophisticated tactics and manoeuvres by both the buyer and the seller. Normally, buyers should not be opportunists and try to function as arbitrators between the factory and the supplier. The buyer must enable his supplier to make a reasonable profit in the short run and survive as well as grow in the long run. He should remember that he is buying not only for today's price, but for tomorrow's cost as well. In this process, the buyers help the suppliers towards finding more economical ways of working rather than squeezing out the profit margin. Selling at a loss will drive the supplier out of business, reducing the sources of suppliers. It is in this context that the art of negotiation is treated as a forum for exchange of views rather than cut-throat buying. In negotiating, both the buyer and seller try to evaluate each other with regard to price and quality.

The parameters affecting the process of negotiation of an item are availability, number of sources, price, delivery, penalty, discounts, package and transportation. The bargaining power will definitely be influenced by the preparedness of the buyer as well as the place and time of negotiation. It is not uncommon to find negotiations taking place over a cup of tea or beer. The known techniques used for negotiation are price analysis, persuasion, discussion, interrogation, investigation, staging a walk-out, prolonged silence, weighing pros and cons, offensive strategy, defensive strategy, blow hot and cold, suggesting complicated formula and learning curve. A successful negotiator has to possess the qualities of patience, persistence, persuasiveness, clear thinking, logical analysis, optimism, knack of getting along with people, ability to plan and be "thick skinned." While negotiating, the buyer should hope for the best and at the same time be prepared for the worst. He should not adopt a strategy of hard choices and soft options. Care should, however, be taken to adopt negotiations on a selective basis as the seller will easily identify the individuals as Mr. five per cent or Mr. ten per cent and inflate the prices accordingly.

PURCHASE BUDGET

Since purchasing activity accounts for a substantial portion of the corporate finance, it assumes very great importance among various budgets such as sales budget, personnel budget and revenue budget. The purchasing budget indicates the purchases to be made for achieving the complete budget plan. These represent the requirements of direct and indirect material and purchased services as set out in the production cost and capital expenditure budgets. The budgets are adjusted in respect of any planned increase or reduction of inventories of raw materials or other supplies including stores and spare parts as well as purchase orders already placed. As far as possible, the requirements should be expressed in physical as well as fiscal terms.

The purchase budget enables the purchase department to plan its purchases and place long-term contracts after considering all relevant factors. It also facilitates the planning of cash requirements. In the budgetary process, periodic statements should be prepared to enable the management to control, and compare the budget with the actual expenditure.

BILL MARKET SCHEME

The discussion on the purchase budget will not be complete without reference to the bill market scheme periodically announced by the Reserve Bank of India. The scheme encourages increasing use of bills of exchange as an instrument of credit. From the point of view of the short-term borrower, the advantage would be that a clean bill of exchange can be discounted at rates lower than those charged by banks on cash credits or overdrafts. The authorities hope that the scheme will not only impose financial discipline on the purchaser but also help the suppliers and producers to plan their financial commitments realistically. Depending upon the "health" of the Indian economy, several changes in the scope of the bill have been introduced.

If a buyer is already receiving credit on purchases from the suppliers, he should then request them to draw the bills of exchange for the period of credit received by the buyer on such extended period of credit up to 120 days. If, on the other hand, the buyer does not receive credit, then also the suppliers may be asked to draw trade bills, the period of usance corresponding to the period required for conversion of stocks into finished goods.

The format in which the bills of exchange should be drawn for rediscount in the Reserve Bank of India under the bill marketing scheme, is shown below:

To

Name and address of buyer

....

....

....

No ....

Place ....

Date ....

At .... days after date/sight (inclusive of days of grace) pay to us or to our order at .... the sum of rupees .... inclusive of interest at ....

....

Signature of the seller

Note: This bill of exchange has been drawn in connection with the sale of .... (specify the description of goods sold) .... covered by .... (number and date of document) like railway receipt, lorry receipt, invoice no., etc.

We accept this .... day of 1974 .... to revise the bill on maturity.

....

Signature of the buyer

10

Purchase Systems

INTRODUCTION

In organisations, depending on the size and nature of operation, the quantum of purchase varies anywhere between a few thousands of rupees and hundreds of crores of rupees. Naturally, this necessitates formalised systems and procedures for ease of operation and accountability. Formal procedures have to be laid down in initiating purchase, selecting suppliers, placing purchase orders, follow-up, receiving materials and so on. In this chapter we will deal with various systems that are used in a purchase department.

We can classify the systems in the following manner:

1. Pre-purchase system;
2. Ordering system;
3. Post-purchase system.

The salient features in each of the systems mentioned are as follows:

1. Pre-purchase System

Initiating the purchase through requisitions, requirements programmes, selection of suppliers, obtaining quotations and evaluating them, are broadly the pre-purchase activities.

(a) *Requisitions.* The department concerned, in need of a material, usually presents a completed requisition form. A typical requisition form is shown in Exhibit 10.1. Requisition may be made by anyone in the concerned department. However, it has to be countersigned by a senior officer. In any organisation only a limited number of officers are empowered to countersign the requisition as it amounts to authorisation of the expenditure. Purchase department must have the list of such officers so as to check the validity of the purchase requisitions. Normally, there is a delegation of authority in authorising a requisition. This is expressed in terms of the financial limits up to which an officer can authorise a requisition. These details must also be available with the purchase department. It is very important to note that capital equipments cannot be requisitioned in this manner. These decisions are taken normally at board level and they are treated differently for taxation and accounting purposes.

EXHIBIT 10.1
MATERIALS REQUISITION FORM

(a) Department requiring the material Date:
(b) Requisition reference number:
To materials management department
Kindly procure the material mentioned below:

Description of the material	Quantity required	Suggested supplier	Purpose	Approximate date when the material is required

Requisitioned by: Authorised by:

(b) *Travelling requisitions.* This document is widely used for requisitioning items that are required frequently in bulk quantities over a long period. The travelling requisition is prepared as shown in Exhibit 10.2. It travels from the requisitioning department to the purchase department often. During each stage a purchase order is initiated. Factors, such as specifications and supplier details, are written permanently and provisions for entering date, quantity required, names of requisitioner and authoriser are available. This reduces paper work and eases the operation. Standardised clerical systems can be devised and the bulk of the work can thus be efficiently handled. For repeat ordering this is the ideal procedure.

EXHIBIT 10.2
TRAVELLING REQUISITION FORM

Material description:
Part number:
Annual usage:
Suppliers: A
B
C

Requisitioned					*Purchased*				*Received*	
Date	Quantity on hand	Requisition	Requisitioned by	Approved by	Date	Quantity	Supplier	Price	Date	Quantity

(c) *Enquiries.* Many organisations often invite suppliers to quote rates for supply of materials. For this purpose a standard format is used which is similar to a purchase order in all respects except that words such as "this is only a request for quotation" or "this is not a purchase order" are printed so as to ensure that the supplier does not construe the request for quotation as a firm order.

2. Ordering System

Having selected the supplier and the rates agreed, the buyer places the purchase order on the supplier, expressing the terms and conditions. The purchase order once accepted becomes a binding contract. The details that are normally furnished in a purchase order are listed below:

(a) Purchase order reference number (which will be quoted in all subsequent follow-up measures pertaining to that order),
(b) Description of the materials and detailed specification,
(c) Quantity required and delivery schedule,
(d) Price and discounts,
(e) Shipping instructions,
(f) Location where the materials are to be shipped (usually the name and address of the buyer),
(g) Signature of the Materials Manager who can authorise the purchase order, and
(h) Detailed terms and conditions (as a common practice these are printed at the back of the purchase order).

By and large, organisations send the supplier an acknowledgement copy along with the purchase order and thereby try to obtain a written acceptance from the supplier. Normally, five or six copies are prepared. The original is sent to the supplier with an acknowledgement copy, which is expected to come back to the materials management department for follow-up and attending to queries. One copy is sent to the receiving department intimating when the consignment is expected so as to facilitate identification. Another copy goes to the indentor for information. The last copy goes to the finance department for subsequent matching with the invoice of the supplier and goods received notes for payment. Depending upon the nature of the organisation, the number of copies required could be increased or reduced.

3. Post-purchase System

This includes follow-up procedures, receipt and checking invoices.

(a) *Follow-up procedures.* Follow-up implies commitment of time and money and therefore it has to be selective. Certain priorities could be established for follow-up. Only 'critical' items require continuous follow-up. Depending upon the movement, items can be classified into fast moving and slow moving. For fast moving items, follow-up can be initiated whenever the stock level depletes to one month's consumption. For slow moving items also similar norms can be established. It is to be noted that these are exceptional follow-up measures. Territorywise field expediting is also resorted to by many organisations. Here, few buyers are specifically made responsible for orders placed in a particular territory and they are based in respective branch offices. The buyer at the branch keeps the headquarters and the plant posted with schedules, problems and despatches. A few organisations also have decentralised inspection facilities for on-the-spot inspection and acceptance. Purchase Order (P.O.) status reports are generally prepared so that selective follow-up is ensured. The status report can be printed part numberwise or supplierwise for suitable follow-up.

(b) *Receipt.* A systematic record of the consignments received, carrier details and descriptions are to be maintained in chronological sequence, to help in quick identification of materials so that inspection can be arranged prior to acceptance. Many organisations have a separate central receiving section for this purpose. As mentioned earlier, a copy of the purchase order is sent to the central receiving section for reconciling purposes.

(c) *Invoice checking.* The supplier normally sends the invoice for the materials supplied for payment. It is essential that this invoice is matched against the receipt details, quantity accepted and rejected so that payments can be made within the discount period or provisions be made which will keep in funds planning. Normally, invoices are sent to the buyer's finance department. A close coordination between the finance and materials management departments is necessary.

SPECIAL PURCHASING SYSTEMS

1. Forward Buying

Forward Buying or committing an organisation far into the future, usually for a year, is in vogue. Depending upon the availability of the item, the financial policies, the economic order quantity, the quantitative discounts, and the staggered delivery, the future commitment is decided. This type of forward buying is different from speculative buying where the motive is to make capital out of the price changes, by selling the purchased items. Manufacturing organisations normally do not indulge in such buying. However, a few organisations do "hedge", particularly in the commodity market by selling or buying contracts.

2. Tender Buying

In public buying, all semblance of favouritism, patronage and personal preferences should be avoided. As such, it is common for government departments and public sector undertakings to purchase through tenders. Private sector organisation adopt tender buying if the value of purchases is more than the prescribed limits as Rs. 50,000 or Rs. 100,000. The steps involved are to establish a bidders' list, solicit bids, evaluate bids by comparing quotations, and place the order with the lowest bidder. However, care has to be taken that the lowest bidder is a responsible party and is capable of meeting the delivery schedule and quality requirement. Open tender system or advertising the tender in newspapers is common in public sector organisations. As advertising bids is costly and time-consuming, most private sector organisations solicit tenders only from renowned suppliers capable of supplying the material.

3. Use of Standard Deviation for Tendering

The use of "Standard Deviation"—explained under Inventory Systems, Chap. 31—for tendering can best be explained by an example. Andhra Electricals Ltd. give the minimum quotation for six lakh incandescent bulbs required in Karnataka Collieries. The bulb manufacturing industry is extremely competitive and hence it is well-known that the lowest bidder will most probably get the tender. The collieries materials manager has announced a bonus scheme—20 paise for each bulb lasting more than 600 hours. This policy is adopted because it will be advantageous to have group replacement for all bulbs lasting more than 600 hours. The director-in-charge of the plant comes to know that it costs a rupee for manufacturing each bulb and Rs 50,000 is the minimum amount for profit and overheads. What should the quotation be for the six lakh bulbs? The laboratory-in-charge of Andhra Electricals reports that the average life of a bulb is 540 hours with standard deviation of 55 hours.

Solution

Manufacturing cost of bulbs	Rs 600,000
Profit and overheads	Rs 50,000
Mean life of bulbs	540 hours
Standard deviation of bulbs	55 hours

The problem is to estimate the proportion of bulbs beyond 600 hours:

$$\frac{600 - 540}{55} = 1.1 \text{ SD}$$

Area for 1.1 SD is .1357 (from Normal tables—Exhibit 31.6)

$$\text{Bonus amount is } \frac{600{,}000 \times .1357 \times 1}{4} = \text{Rs } 20{,}355$$

The Quotation is 650,000 – 20,355 = Rs 629,645.

Similar methods can be used for evaluating the bids and choose the supplier.

4. Blanket Order

This system minimises the administrative expenses and is useful for "C" items. It is an agreement to provide a required quantity of specified items, over a period of time, usually for one year, at an agreed price. Deliveries are made depending upon the buyer's needs. The system relieves the buyer from routine work, giving him more time for focussing attention on high value items. It requires fewer purchase orders and thus reduces clerical work. It often achieves lower prices through quantity discounts by grouping the requirements. The supplier, under the system, maintains adequate inventory to meet the blanket orders, but he does not incur selling costs, once the negotiations are finalised.

5. Zero Stock

Some firms try to operate on the basis of zero stock and the supplier holds the stock for these firms. Usually, the firms of the buyer and seller are close to each other so that the raw material of one is the finished product of another, as for instance, naptha in a fertiliser plant is one of the final products of the refinery. Alternatively, the system could work well if the seller holds the inventory and if the two parties work in close coordination. However the price per item in this system will be slightly higher as the supplier will include the inventory carrying cost in the price. In this system, the buyer need not lock up the capital and so the purchasing routine is reduced. This also significantly reduces obsolescence of inventory, lead time and clerical efforts in paper work. Thus, the seller can devote his marketing efforts to other customers and production scheduling becomes easy.

6. Rate Contract

The system of rate contract is prevalent in public sector organisations and government departments. It is common for the suppliers to advertise that they are on "rate contract" with the DGS&D, for the specific period. After negotiation, the seller and the buyer agree to the rates of items. Application of rate contract has helped many organisations to cut down the internal administrative lead time as individual firms need not go through the central purchasing departments and can place orders directly with the suppliers. However, suppliers always demand higher prices for prompt delivery, as rate contracts normally stipulate only the rate and not the delivery schedule. This difficulty has been avoided by ensuring the delivery of a minimum quantity at the agreed rates. This procedure of fixing a minimum quantity is called the running contract and is being practised by the railways and the DGS&D.

The buyer also has an option of increasing the quantity by 25 per cent more than the agreed quantity under this procedure.

7. Reciprocity

Reciprocal buying means purchasing from one's customers in preference to others. It is based on the principle "if you kill my cat, I will kill your dog", and "Do unto your customers as you would have them do unto you". Other things, like soundness from the ethics and economics point of view being equal, the principles of reciprocity can be practised. However, a purchasing executive should not indulge in reciprocity on his initiative when the terms and conditions are not equal with other suppliers. It is often found that less efficient manufacturers and distributors gain by reciprocity what they are unable to gain by price and quality. Since this tends to discourage competition and might lead to higher prices and fewer suppliers, reciprocity should be practised on a selective basis. We have dealt with this aspect in the Chapter on "Buyer-Seller Relations" also.

8. Systems Contract

This is a procedure intended to help the buyer and the seller to reduce administrative expenses and at the same time ensure suitable controls. In this system, the original indent, duly approved by competent authorities, is shipped back with the items and avoids the usual documents like purchase orders, materials requisitions, expediting letters and acknowledgements, goods in words report, etc. The contract is invariably simple, covering only delivery period, price and invoicing procedure. Carborandum Company in the US claims drastic reduction in inventory and elimination of 40,000 purchase orders by adopting the systems contracting procedure.* Systems contracting is particularly useful for items with low unit price with high consumption profile and relieves the buyers of the routine work.

*Bolton, R.A., "System Contracting: A New Way to Reduce Purchasing Costs" *Management Review*, May 1967.

11

Price Forecasting

IMPORTANCE

Forecasting is essential for sound functioning as well as the survival of any firm. Through forecast of demand, we evaluate the factors in the environment which influence the operations of the firm and formulate annual plans. In materials management, too, forecasting is a vital and inescapable aspect. We forecast the requirements of various materials and their prices, and based on these forecasts we formulate the materials budget, cashflows, inventory levels, etc. Although in this chapter we will be concerned with price forecasting, the techniques used here can be applied for demand forecasting also, with suitable modifications.

Forecasting techniques discussed below will try to reduce the subjectivity; however, one can consider all factors which influence the price. Some of these factors cannot be quantified and Delphi techniques discussed in Chap. 32 can be used to take care of such intangible factors.

All forecasting techniques depend on the manner in which the basic data is collected, evaluated and used for their reliability. The data collection depends upon factors such as classification, reliability, accuracy, source, end use and cost. It also depends upon whether the data is required for programmed or non-programmed decisions.

APPROACHES

Many techniques are available in forecasting. Each firm will have to carefully assess its data processing facilities, skills and the operating environment so as to select a technique that is most suitable. In inventory systems, factors which influence forecasting are the number of items involved, the nature of their markets, and the data handling system. Techniques which are normally used are moving averages methods, exponential smoothing, regression analysis, etc. Sometimes Delphi technique is used for long-term forecasts. Basic issues, such as forecasting technique, forecast time detail, and review frequency must be finalised in consultation with top management.

PRICE FORECASTING TECHNIQUES

One of the important activities in materials management is the forecasting of prices of materials to be procured. The forecast of material prices forms the basis of purchase budgeting, product costing, stocking policies, etc.

Materials Research provides the basic infrastructure and information for the forecast of prices. The research activity helps in predicting some sudden changes in prices due to environmental causes. Executive brain-storming sessions can also help anticipate sudden price changes. However, even without any observable environmental cause, prices do rise and they need to be forecasted on the basis of statistical analysis. This statistical analysis for price forecasting is valid only for random changes in prices and any observable causes for their variations have to be accounted for by applying suitable correction factors to the statistical forecast.

THUMB RULE FOR PRICE FORECASTING

A general practice in many organisations is to use a 10 per cent annual rise in price as a basis for forecast. This is all-right for an economy in the grip of inflation. However, the prices of some products vary randomly both within one year and between years and the prices of these can be predicted by using statistical methods.

The forecasting techniques can be broadly classified as those based on (i) the extrapolation of the same price series, and (ii) relation with another price series.

EXTRAPOLATION OF THE SAME PRICE SERIES (Time series)

The following two methods conform to the above title: (i) Moving average method, and (ii) Exponential smoothing. These are explained by means of an example.

Consider the prices of linseed oil in the years 1970 and 1971. The monthly prices of linseed oil per quintal are indicated below:

EXHIBIT 11.1

Month	*1970*	*1971*
	(Rs per quintal)	
January	454.0	481.0
February	445.0	456.2
March	440.0	439.5
April	384.0	413.8
May	396.0	406.0
June	414.0	393.8
July	404.0	389.0
August	405.0	401.7
September	432.5	421.7
October	422.0	425.0
November	464.0	432.5
December	469.0	412.3

By using the five month moving average method, the prices from June 1970 up to December 1971 could have been predicted. The predicted value for June 1970 would be the sum of the actual prices for the months January, February, March, April and May divided by 5. The predicted price value for July 1970 would be the sum of the actual prices of February, March, April, May and June divided by 5. Hence, the predicted price for June 1970 would be: $\frac{454 + 445 + 440 + 384 + 396}{5}$ which is Rs 423.80 per quintal and the predicted value for July would be $\frac{445 + 440 + 384 + 396 + 414}{5}$ which is Rs 415.80 per quintal. Proceeding in the same manner, the predicted prices for the other months could be worked out. This is shown in Exhibit 11.2.

EXHIBIT 11.2

	Predicted price	*Actual price*
1970		
June	423.80	414.00
July	415.80	404.00
August	407.60	405.00
September	400.60	432.50
October	410.30	422.00
November	415.50	464.00
December	425.50	469.00
1971		
January	438.50	481.00
February	453.70	456.20
March	458.44	439.50
April	461.94	413.80
May	451.90	406.00
June	439.70	393.80
July	422.26	389.00
August	408.82	401.70
September	401.26	421.70
October	403.84	425.00
November	407.64	432.50
December	415.38	412.30

In the moving average method, the weightages given to the price of each year is $1/n$ where n is the number of months over which the average is taken. In the illustrative example, the number of months was 5 and the weightage given to each of the last 5 months' prices was 1/5 or 0.2.

The moving average method requires a lot of data preservation and this becomes tedious if the number of months over which the average is taken is large. Another aspect for consideration is the equal weightages given to all the previous month's prices. These two aspects, viz. data preservation and giving equal weights to previous months' prices, is overcome by using what is known as an Exponential Smoothing Method.

In the Exponential Smoothing Method, the forecast for the next month is based on the forecast for the previous month and the actual price of the previous month. The weightages given are dependent on the nature of changes in prices.

$$\text{Forecast of month } (m) = \frac{1}{n} \text{ [Actual price in month } (m - D)$$

$$+ \left(1 - \frac{1}{n}\right) \text{Forecast for month } (m - D)].$$

If $n = 5$ (n has to be chosen based on experience) then the forecast for the month July 1970 = 0.2 (414) + 0.8 (423.80) = 82.80 + 339.04 = Rs 421.84 per quintal.

The Forecast price for August 1970 would be 0.2 (404) + 0.8 (421.84) = 80.80 + 337.47 = 418.27/quintal.

The Forecast values and the actual prices up to December 1971 are shown in the table below:

EXHIBIT 11.3

1970	*Predicted*	*Actual*	*1971*	*Predicted*	*Actual*
June	423.80	414.00	April	444.52	413.80
July	421.84	404.00	May	438.38	406.00
August	418.27	405.00	June	431.90	393.80
September	415.31	432.50	July	424.28	389.00
October	418.75	422.00	August	416.22	401.70
November	419.40	464.00	September	413.32	421.20
December	427.32	469.00	October	414.00	425.00
1971					
January	435.66	481.00	November	416.20	432.50
February	444.73	456.20	December	419.46	412.30
March	447.02	439.50			

In the above illustration the weightage given to previous months' price is 0.2. The prior months get weightages (0.2 × 0.8), (0.2 × 0.8 × 0.8); and so on. The weightages given to months is inversely related to their age. For the month June 1971, the weightage given to May 1971 price is 0.2, the weightage to April is 0.16, to March 1971 is 0.128 and so on.

This is explained below:

(Forecast for June 1971) = Actual of May 1971 × 0.2 + 0.8 (Forecast for May 1971)

= Actual of May 1971 × 0.2 + 0.8(0.2) Actual of April 1971 + 0.8 (Forecast for April 1971)

= Actual of May 1971 × 0.2 + Actual of April 1971 × 0.16 + 0.8 (0.8) 0.2 (Actual of March 1971) + 0.8 (Forecast for May 1971)

= 0.2 (Actual of May 1971) + 0.16 (Actual of April 1971) + 0.128 (Actual of March 1971) + 0.512 (Forecast for March 1971).

The weightages attached to previous months' prices for June 1971 forecast is indicated graphically below:

RELATION WITH ANOTHER PRICE SERIES

The linseed oil price is based on the price of raw linseed. The linseed oil price is derived from the linseed price by a forecasting technique known as Regression. The linseed oil price is then forecast on the basis of the raw linseed price forecast and the linear relationship between raw linseed price and linseed oil price.

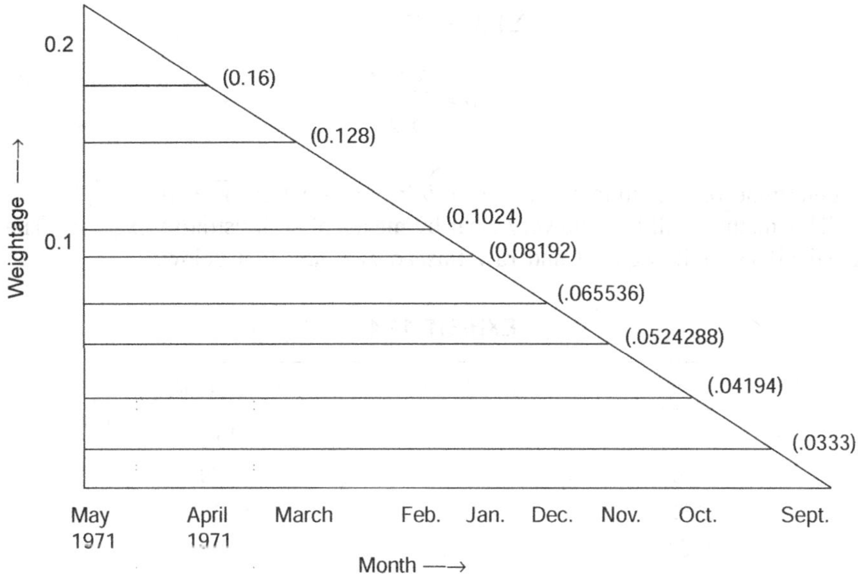

If Y_t represents linseed oil price in any year t and X_t, the raw linseed price in the same year, then

$$Y_t = a + bX_t.$$

If a and b are determined by regression analysis, the linseed oil price can be forecast based on the forecast for raw linseed.

For values of $t = 1$ to n

(1) $Y_1 = a + bX_1$

(2) $Y_2 = a + bX_2$

$\vdots$

(n) $Y_n = a + bX_n$

Adding all the n equations

$$(n+1): Y_1 + Y_2 + \dots + Y_n = na + b(X_1 + X_2 + \dots + X_n)$$

Dividing throughout by n

$$\frac{\sum Y_t}{n} = a + b\frac{\sum X_t}{n}$$

$(n + 2)$ or $\overline{Y}$ (average value of Y_t) $= a + b(\overline{X})$.

Subtracting equation $(n + 2)$ from any of the first n equations $(Y_t - \overline{Y}) = b(X_t - \overline{X})$, i.e., the difference of Y_t from the average = b(the difference of X_t from average), or

$$Y_t = b(X_t)$$

$$Y_t X_t = b(X_t)^2$$

$$\Sigma Y_t X_t = b\Sigma X_t^2$$

$$b = \frac{\Sigma Y_t X_t}{\Sigma X_t^2}$$

To determine a, substitute the value of b in the equation $\overline{Y} = a + b(\overline{X})$.

This method will become very clear by means of an illustrative example. The annual prices of linseed oil and raw linseed are indicated below.

EXHIBIT 11.4

Year	*Raw linseed*	*Linseed oil*
	(X_t)	(Y_t)
1961	78.86	117.39
1962	79.88	183.15
1963	73.22	158.58
1964	94.47	213.42
1965	119.24	250.75
1966	158.05	361.54
1967	188.45	404.50
1968	130.37	280.64

$$\overline{X} = \frac{\Sigma X_t}{8} = \frac{922.54}{8} = 115.3$$

$$\overline{Y} = \frac{\Sigma Y_t}{8} = \frac{1969.97}{8} = 246.25$$

Y_t	X_t	Year	Y_tX_t	X_t^2
(–128.86)	(–36.44)	1961	4695.66	1327.87
(–63.10)	(–35.42)	1962	2235	1254.78
(–87.67)	(–42.08)	1963	3689.15	1770.73
(–32.83)	(–20.83)	1964	683.85	433.89
(+4.50)	(+3.94)	1965	17.73	15.52
(+115.29)	(+42.75)	1966	4928.65	1827.56
(+34.39)	(+15.07)	1968	518.26	227.10

$$\Sigma Y_tX_t = 28{,}344.29, \quad \Sigma X_t^2 = 12{,}208.37$$

$$\therefore \qquad b = \frac{\Sigma Y_tX_t}{\Sigma X_t^2} = 2.322$$

$$\overline{Y} = a + b\overline{X}$$

or

$$246.25 = a + (2.322)\ 115.30$$

$$= a + 267.75$$

$$a = 246.25 - 267.75$$

$$= -21.50$$

$\therefore$ Linseed oil price = 2.322 (Raw linseed price) – 21.50

The actual prices are shown against those predicted by regression analysis below:

Year	*Actual value*	*Predicted value*
1961	117.39	161.60
1962	183.15	164.00
1963	158.48	148.50
1964	213.42	197.90
1965	250.75	255.40
1966	361.54	345.50
1967	404.50	416.10
1968	280.64	281.20

CONTROL

We have discussed the importance of forecasts in relation to materials management, various techniques of forecasting and their applicability. Setting up a reliable forecasting system is undoubtedly a very important job. However, it will be incomplete if there is no formal procedure for their review. This is the control aspect where the forecasting system is reviewed for improvements in effectiveness. Use of forecasting techniques in the framework discussed earlier can give very good results.

12

Buying Seasonal Commodities

In a free market under perfect conditions the prices would change as dictated by the disparity between demand and supply. It is, therefore, necessary to examine the seasonal behaviour of demand and supply.

NATURE OF SEASONALITY

Seasonality could be natural or man-made. Natural seasonality is most common in the agricultural industry. Man-made seasonality mainly arises because the operations of an organisation are reflected in the account books. The purchasing and sales activity will be at its peak as the financial year nears its end. This is because the yearly sales targets have to be met and the purchase budgets exhausted.

The seasonal aspects of demand and supply could be for independent reasons. This is indicated below:

1. *Supply.* Agricultural-based materials such as foodgrains, linseed oil, groundnut oil, vegetable oil, etc. The demand for these products is, however, uniform during a year.

2. *Demand.* Agricultural implements, fertilisers, paints, construction material, etc. Paints and construction materials are treated as seasonal demand as their sales are low during the monsoons. The supply of these products is, however, uniform.

The purchase executive is not concerned with a situation of seasonal demand and uniform supply. He is faced with a peculiar problem when the supply is seasonal and the demand uniform. The problem faced is two-phased: (i) to assure continuous supply to the company, and (ii) to make purchases at a reasonable price, since the prices fluctuate.

In the case of essential foodgrains and imported raw materials like non-ferrous metals, two public sector undertakings—the Food Corporation of India and the Minerals and Metals Trading Corporation—have been set up to assure continuous supply. These organisations naturally indulge in forward buying in order to effect continuous supply and in the process are restricted by two major factors: (i) the available warehousing facility, and (ii) the amount of liquid capital at hand.

The Mineral and Metals Trading Corporation procures non-ferrous metals when the world market prices are low. Similarly, the purchase executives relying on linseed oil, vegetable oil or rubber (as raw materials) serve their organisations by procuring at low prices. This aspect of purchasing under fluctuating prices needs a very methodical approach.

PURCHASING UNDER FLUCTUATING PRICES

If the buyer purchases at regular intervals to meet the requirements of the factory, he will sometimes pay a high or a low price, but in the long range the overall cost will be equal to the average market price. Such a "conservative policy" is inevitable if the factory has no facilities for storing in excess of requirements and the buyer chooses not to buy forward. However, if storage space is available, the buyer can, to a limited extent, buy more in anticipation of high prices.

To do this successfully, requires both skill and a certain amount of luck. It is necessary to balance the probability of the price falling in the near future against the cost of buying now and storing for later use. Inevitably, the ability to forecast future prices with accuracy plays an important part, but the fact remains that no forecast can be accurate. Therefore, rather than relying on fallible human judgements, a technique has been developed which ensures that the average price paid is as low as possible. The name of the technique is Dynamic Programming. In practice, the optimal price paid resulting from the use of Dynamic Programming will be greater than the cost of buying with perfect hindsight—the difference represents the cost of uncertainty, the cost of being unable to predict the future accurately.

To illustrate the technique let us consider the following example:

A company requires a certain type of vegetable oil at the rate of 100 tons per week. The storage space available is for an absolute maximum of 1000 tons, representing ten weeks' demand. It is forecast that the price of this particular oil will vary randomly between Rs 5,000 per ton and Rs 6,000 per ton in the foreseeable future.

The company requires a rule to determine how much oil to buy any week knowing the prevailing price and the amount already in stock. This rule is to be determined so that in the long run the company pays the minimum amount for the oil. The market prices forecast over a period of next 50 weeks are given below.

EXHIBIT 12.1

Weeks	*Weekly prices** (Rs, ton)				
1 5	5,770	5,650	5,200	5,090	5,800
6–10	5,340	5,460	5,020	5,050	5,040
11–15	5,150	5,400	5,060	5,870	5,880
16–20	5,170	5,190	5,770	5,660	5,140
21–25	5,680	5,270	5,850	5,110	5,170
26–30	5,260	5,950	5,680	5,970	5,730
31–35	5,750	5,640	5,270	5,450	5,010
36–40	5,870	5,200	5,020	5,190	5,260
41–45	5,450	5,420	5,960	5,720	5,980
46–50	5,780	5,810	5,520	5,310	5,880

*The prices are plotted in Exhibit 12.2.

The conservative policy. If the company were to go to the market and buy 100 tons, regardless of the market price, it would buy 5,000 tons over a period of 50 weeks at an average price of Rs 5,480 per ton. This is the price the company would have to pay if there were no storage price that the company need pay in the long run.

Hindsight policy. If the forecasts of prices were perfect (no variance at all), the company could buy at the lowest possible price in the long run. All it needs to do is to look at the prices in the next 10 weeks.

Starting with week 1 (assume no initial stock) the company must buy at least 100 tons at Rs 5,770 per ton. No more than this should be bought since the price falls in the next week to Rs 5,650/ton.

Week no.	*Opening stock*	*Price paid*	*Amount bought*	*Closing stock*
1	Nil	5,770	100	Nil

The same argument holds good for weeks 2 and 3, but in week 4 the price of Rs 5,000 is the lowest until week 8 and hence the company should buy 400 tons to keep supplies until week 8.

Week no.	*Opening stock*	*Price paid*	*Amount bought*	*Closing stock*
1	Nil	5,770	100	Nil
2	Nil	5,650	100	Nil
3	Nil	5,200	100	Nil
4	Nil	5,090	400	300

In week number 8 the price is Rs 5,020/ton, lower than at any time for the next ten weeks and so the maximum quantity (1,000 tons) that can be stored is purchased.

Week no.	*Opening stock*	*Price paid*	*Amount bought*	*Closing stock*
1	Nil	5,770	100	Nil
2	Nil	5,650	100	Nil
3	Nil	5,200	100	Nil
4	Nil	5,000	400	300
8	Nil	5,020	1,000	900

In week 9 although the price is low (Rs 5,050) it is not the lowest in the next ten weeks—in fact it is lower in week 10 (Rs 5,040) and so the company should hold on until then and buy the maximum amount that they can stock, viz., 200 tons.

Week no.	Opening stock	Price paid	Amount bought	Closing stock
1	Nil	5,770	100	Nil
2	Nil	5,650	100	Nil
3	Nil	5,200	100	Nil
4	Nil	5,090	100	300
8	Nil	5,020	1,000	900
10	800	5,040	200	900

And thus the company proceeds each week by buying as much as necessary to keep it running until a lower price can be obtained. The full result of such buying is shown in the following table.

Complete Buying Record Using Hindsight

Week no.	Opening stock	Price paid	Amount bought	Total expenses
1	Nil	5,770	100	5,77,000
2	Nil	5,650	100	5,65,000
3	Nil	5,200	100	5,20,000
4	Nil	5,090	400	20,36,000
8	Nil	5,020	1,000	50,20,000
10	800	5,040	200	10,08,000
13	700	5,060	300	15,18,000
20	300	5,140	100	5,14,000
24	Nil	5,110	1,000	51,10,000
25	900	5,170	100	5,17,000
35	Nil	5,010	1,000	50,10,000
38	700	5,020	300	15,06,000
39	900	5,190	100	5,19,000
49	Nil	5,210	200	10,62,000
			5,000	2,54,82,000

$$\text{Average price paid} = \frac{2{,}54{,}82{,}000}{5{,}000} = 5{,}096.40/\text{ton}.$$

The average price paid now is Rs 5,096.40/ton as compared to Rs 5,480/ton when no storage space was available. The problem is how to reduce the figure of Rs 5,480/ton without accurate forecasts of prices.

DETERMINING THE OPTIMAL BUYING STRATEGY

In building up this optimal buying strategy (as also in case of conservative and hindsight buying) the following assumptions are implicit.

1. Time has been divided into successive periods; in this example, weeks.
2. A "planning horizon" of N periods is arbitrarily taken. In this example, a storage capacity of 10 periods (weeks) is imposed. Note that even with

infinite storage capacity a company cannot have infinite periods as the knowledge of prices will be diffused after a certain number of periods.

3. Demand is known for all the periods in the planning horizon. In this example, a uniform demand of 100 tons per period is assumed. It can be different for different periods in other situations.
4. No shortages are allowed. Whatever be the price a company has to buy (if it does not have in stock) at least one period's requirement.
5. No storage cost has been assumed. If, however, the "planning horizons" are longer, the storage cost has to be considered.
6. There is only one price offer in a period. If this opportunity of purchase is foregone, a company has to wait for the next period to purchase its requirements.
7. There are no supply restrictions. Quantities up to the entire requirement of "planning horizon" are available in the market during any one of the periods.
8. The prices fluctuate randomly and are independent of each other, at least during the "planning horizon".
9. Delivery is assumed to be instantaneous.
10. Stock once bought cannot be resold.

It is imperative that a company should hold off when the prices are high and buy when the prices are low. Such a decision is very much dependent on the stock on hand. If the stock on hand is low, a company might decide to buy even at reasonably high prices and if it is high it might not buy at relatively lower prices. The optimal technique that a company should use is based on the prices of indifference, i.e., given the stock on hand, there would be a price when the company would be indifferent between buying and holding off. The nature of such prices of indifference will be as follows:

EXHIBIT 12.2

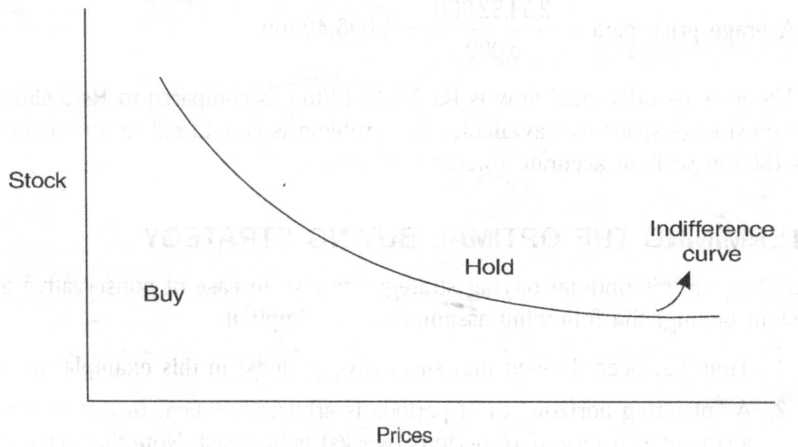

Having plotted the price of indifference curve a company has only to go to the market, check up the price and raise the stock to the level of indifference.

Thus, if P_0, P_1, ..., P_9 are the prices of indifference when stocks are for zero weeks, 1 week ... 9 weeks, what a company needs is to plot these prices against the stock levels.

Calculation of P_0. If a company has no stock on hand, it has to buy at least one week's requirement no matter what the price is. Since the highest price is Rs. 6,000/ton, P_0 = Rs 6,000.

Calculation of P_1. The company has one week's stock on hand. It has to decide whether it should buy or hold off. We are essentially looking at what the price would be next week. The prices vary randomly (assume prices to be uniformly distributed between Rs 5,000 and Rs 6,000 per ton) and the expected price next week would be Rs 5,500. Therefore, P_1 = Rs 5,500, i.e., buy with one week's stock on hand if the price is below Rs 5,500 per ton, say, buy the next week's requirement as well in this week if the price is below Rs 6,500 per ton.

Calculation of P_2. The company has stock for two weeks. It has to decide whether to buy the third week's requirement now or hold off. The possible ways of buying the third week's requirements are:

EXHIBIT 12.3

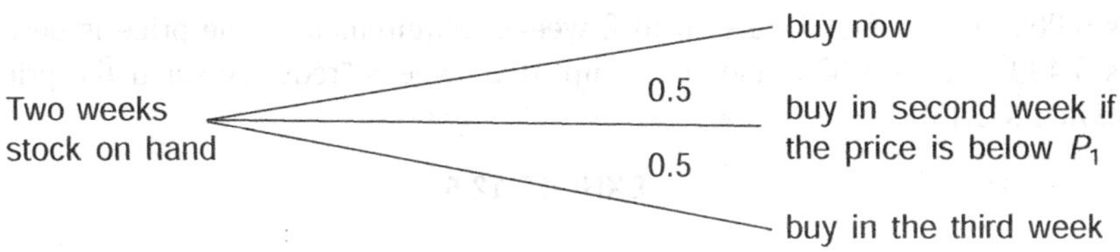

Suppose the decision is to hold off. Next week (i.e., second week), the company will buy if the price is below P_1, i.e., Rs 5,500. The chances of the price being below Rs 5,500 is 0.5 (50% of the times) and its expected value is Rs 5,250. There is again 50% chances of the price in the second week being above Rs 5,500, and in such a case the company will buy in the third week. The expected price in the third week is Rs 5,500.

Therefore

$$P_2 \text{ (price of indifference)} = 0.50\ (5250) + 0.50\ (5500)$$
$$= 2625 + 2750 = \text{Rs } 5{,}375$$

i.e., if the price this week is below Rs 5,375/ton, buy the third week's requirement now.

Calculation of P_3. The company has 3 weeks' stock on hand. It has to decide whether or not to buy the fourth week's requirement. The alternatives are:

EXHIBIT 12.4

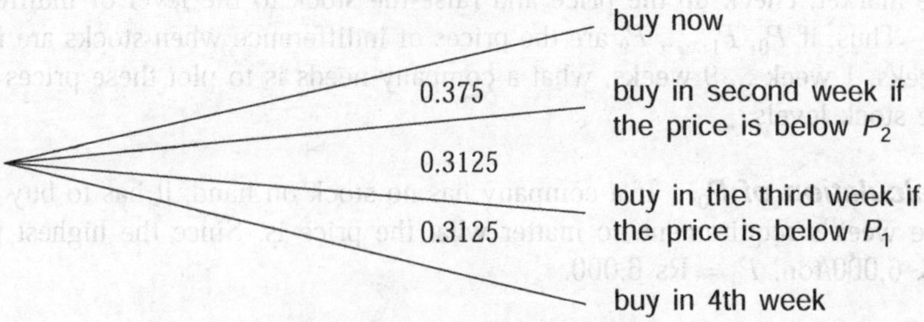

Therefore,

$$P_3 = 0.375\ (5187.50) + 0.625\ (0.5)\ (5250) + 0.625\ (0.5)\ (5500) = \text{Rs } 5{,}305.$$

Proceeding along the same lines the company can calculate all the prices of indifference up to P_9. The values of these prices are:

P_1 = Rs 5,500, P_2 = Rs 5,375, P_3 = Rs 5,305, P_4 = Rs 5,258, P_5 = Rs 5,225

P_6 = Rs 5,200, P_7 = Rs 5,180, P_8 = Rs 5,160, P_9 = Rs 5,150

This means: raise stock to one week's requirement if the price is between Rs 6,000 and Rs 5,500, raise it to 2 weeks' requirement if the price is between Rs 5,499 and Rs 5,376, and so on, up to 10 weeks' requirement if the price is *below* Rs 5,150.

EXHIBIT 12.5

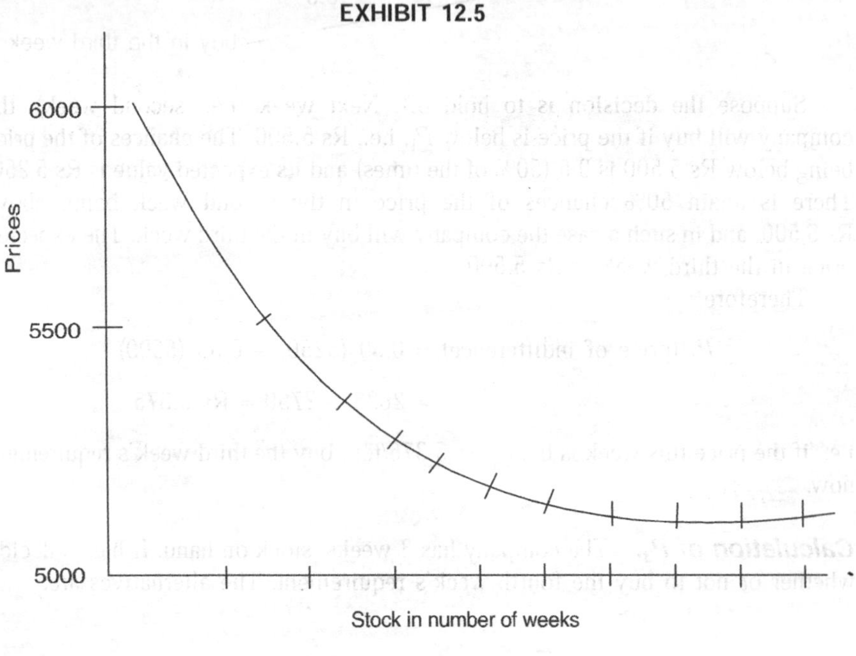

The company can estimate the average price paid in the long run by applying these prices of indifference to the prices of the 50 weeks which is forecast.

In week 1, the price is Rs 5,770; therefore, buy only one week's requirement.

Week no.	*Opening stock*	*Price*	*Amount bought*	*Closing stock*
I	Nil	5,770	100	Nil

In week 2, the price is Rs 5,650, therefore the company should buy a week's requirement.

Week no.	*Opening stock*	*Price*	*Amount bought*	*Closing stock*
1	Nil	5,770	100	Nil
2	Nil	5,650	100	Nil

In week 3, the price is Rs 5,200. This is equal to P_6—the company should raise the stock to 6 weeks' requirement. (Note if the price is below Rs 5,200, the company would buy a week's requirement even if the stock on hand was six weeks' requirement).

Week no.	*Opening stock*	*Price*	*Amount bought*	*Closing stock*
1	Nil	5,770	100	Nil
2	Nil	5,650	100	Nil
3	Nil	5,200	600	500

In week 4, the price is Rs 5,090—this is even below P_9; hence the company should raise the stock to full capacity.

Week no.	*Opening stock*	*Price*	*Amount bought*	*Closing stock*
1	Nil	5,770	100	Nil
2	Nil	5,650	100	Nil
3	Nil	5,200	600	500
4	500	5,090	500	900

Proceeding on these lines for all the 50 weeks, the amount purchased and price paid for each of the purchases made is listed below:

Week no.	Opening stock	Amount bought	Price	Total expenses
1	Nil	100	5,770	5,77,000
2	Nil	100	5,650	5,65,000
3	Nil	600	5,200	31,20,000
4	500	500	5,090	25,45,000
8	600	400	5,020	20,08,000
9	900	100	5,050	5,05,000
10	900	100	5,040	5,04,000
13	700	300	5,060	15,18,000
16	700	100	5,170	5,17,000
20	400	600	5,140	30,84,000
24	600	400	5,110	20,44,000
33	100	300	5,270	15,81,000
35	200	800	5,210	40,08,000
38	700	300	5,020	15,06,000
43	Nil	100	5,520	5,52,000
49	Nil	200	5,310	10,62,000
		5,000		2,56,96,000

$$\text{The average price paid} = \frac{2{,}56{,}96{,}000}{5{,}000} = \text{Rs } 5{,}139.20.$$

The results obtained can now be tabulated as:

Conservative policy = Rs 5,480

Hindsight policy = Rs 5,096.40

Optimal policy = Rs 5,139.20

The difference between Hindsight and Optimal policy, i.e., Rs 42.80 per ton can be viewed as the cost of uncertainty. The annual cost of optimal policy is Rs 2,56,96,000 and that of conservative policy is Rs 2,74,00,000 and the saving is Rs 17,04,000 for 50 weeks or nearly one year.

13

Purchasing Under Uncertainty

CLASSIFICATION OF DECISION PROBLEMS

The primary task of the materials manager is to decide on the acquisition of the necessary raw materials and supplies to support orderly production at the lowest ultimate cost.

The essential features of all purchase decision-making problems are:

1. recognition of the existence of several possible strategies;
2. prediction of the pay-off of each one of the strategies; and
3. assessment of the order of preference of the strategies.

It may be noted in this context that at least for some of the strategies, the amount of profit is unknown because it will be determined by some event which cannot be predicted with certainty.

All decision-making problems can be classified into three categories—certainty, risk, and uncertainty; depending upon the information available to the executive in a particular situation.

UNCERTAINTY SITUATION

Purchasing decision-making under conditions of uncertainty arises when the information of various strategies to be adopted in a case neither logically available nor is it possible to estimate the probability of occurrences of each of these strategies. The situation should not be confused with total ignorance. The executive will surely have some knowledge of the strategies though this knowledge is not sufficient to obtain explicit data on which the calculation of probabilities of different strategies could be based. Such problems are often encountered while buying a new product, procuring from a new source and speculative purchases. Thus, decision-making problems under conditions of uncertainty occur because the information available to the executive is unavoidably inadequate. Several statistical criteria are available to the executive to face any such situation with optimism. The application of these criteria depends upon the attitude of the 'decision-maker' in a problem situation and these are illustrated with the help of the following example:

The necessity for a special kind of new laboratory apparatus is expected to continue either for four years or for five years, but the probabilities cannot be estimated because due to technological advancement, the apparatus will be replaced after four or five years. Two types of apparatuses X and Y are available in India. Each apparatus consists of an important part and this part will need a complete overhaul in either the fourth or fifth year. No data are available which permit the estimation of the probabilities that the overhaul will occur in the fourth or fifth year. The relevant figures of costs in rupees after obtaining pertinent information from the supplier are given in Exhibit 13.1.

EXHIBIT 13.1

Type of cost	*Apparatus X*		*Apparatus Y*	
Capital investment cost	5,000		7,500	
Operating cost per year	1,000		500	
Preventive maintenance expenses	P_1	P_2	P_3	P_4
Year 1	200	200	200	100
Year 2	200	200	100	100
Year 3	300	300	150	150
Year 4	700	300	1000	200
Year 5	200	700	100	1000

Two sets of figures are given in this table for each apparatus depending upon whether the overhauling is done in the fourth or fifth year. Under the two conditions of uncertainty, viz., requirement in years and overhauling period, the problem is to decide on the purchase of one apparatus.

For each apparatus there are four possible alternative corresponding to two periods of use and two periods of overhauling. The average annual costs incurred may be calculated in any case by the following steps. For simplicity, the scrap in this case and other sophisticated costs have been assumed to be zero and due arithmetical adjustments can be made if scrap value is to be considered. This illustration may be suitably extended if there are increased number of apparatuses or any other similar kind of uncertainty situations.

Apparatus X	Life 4 years Overhauling
Investment	Rs 5,000
Total operating cost per 4 years	Rs 4,000
Overhauling expenses 200 + 200 + 300 + 700	Rs 1,400
Total operating cost per 4 years	Rs 10,400
Average expenses per year	Rs 2,600

Similarly, all permutations and combinations can be worked out for the different alternatives and the average annual expenses in rupees can be tabulated as in Exhibit 13.2 (each figure is duplicated in order to complete all comparisons and also in preparing the *regret* matrix explained later).

EXHIBIT 13.2

Life in years		*4*	*5*	*4*	*5*	*4*	*5*	*4*	*5*
Preventive Maintenance	X	p^1	p^1	p^2	p^2	p^1	p^1	p^2	p^2
Preventive Maintenance	Y	p^3	p^3	p^3	p^3	p^4	p^4	p^4	p^4
Apparatus	X	2,600	2,320	2,500	2,340	2,600	2,320	2,500	2,340
Apparatus (Regret matrix concept explained)	Y	2,713	2,290	2,713	2,290	2,513	2,310	2,515	2,310
Apparatus	X	0	30	0	50	87	10	0	30
Apparatus	Y	113	0	213	0	0	0	15	0

After having obtained the given information in the above manner, the following principles help the executive in choosing one apparatus.

Principle of Rationality

The "criterion of Laplace" is useful to an executive when he does not have any preferences in choosing the alternatives in a given decision-making problem. This assumes that all possible strategies are likely to occur with the same chance or with equal subjective probability in the absence of adequate information on them. The decision rule to be followed then is to choose that strategy which maximises the average or expected gain. In the given illustration the average expense for all combinations of X or E (X)—expected value of X—is Rs 2,440 and for Y or E(Y) is Rs 2,456. Since the average expense of X is expected to be less, apparatus X is to be preferred in this situation. This principle is usually adopted by practising managers.

Principle of Minimax

The second criterion which was established by Abraham Wald is known as minimax criterion or criterion of pessimism. This criterion is suited to a problem situation which forces the attitude of the executive to be very conservative, pessimistic, unadventurous and compels him to play too safe in order to protect him from the falling of the Damocles' sword. In such a situation, the decision rule is to examine the minimum profit or maximum cost associated with each strategy and choose that strategy which maximises the minimum profit or minimises the maximum cost so that the worst outcome is made as desirable as possible. It is observed from Table 13.2 that worst X is Rs 2,600, whereas the worst Y is Rs 2,713 and the best out of the two, namely, apparatus X is to be chosen in this particular situation.

Principle of Maximax

It may be noted that on the other extreme end of the minimax rule is the maximax criterion. This deals with a problem situation which forces the attitude of the executive in that situation to be rash and optimistic, thinking that the circumstances are going to be benevolent always. In such a case the executive chooses that strategy which maximises the maximum profit. The best X from Exhibit 13.2 is Rs 2,320, best Y is Rs 2,290 and so apparatus Y is to be preferred under these circumstances.

Principle of Optimum

Hurwicz's criterion of optimism is brought in to avoid extreme conservatism of the mimimax principle or extreme radicalism of the maximax principle. The criterion allows for the varying degrees of moderation according to the variation in optimism or pessimism of the attitude of the decision-maker in a specific problem situation. An index of optimism L is obtained from the executive to determine the coefficient of optimism of the executive in a specific problem situation. It may be noted in this context that if the index of optimism $L = 1$, the criterion reduces to the maximax principle. A knowledge of L will enable one to choose the best strategy in such a specific problem situation. From Exhibit 13.2, the following equations can be obtained. If there is no preference between the two apparatuses, then:

$$2320L + 2600(1 - L) = 2290L + 2713(1 - L), \text{ i.e., } 143L = 113$$

When $L = 0.79$, X or Y may be chosen; when L is less than 0.79, apparatus X is preferred (to coincide with minimax principle; when L is greater than 0.79, apparatus Y is preferred (to coincide with maximax principle).

Principle of Regret

Savage's criterion of regret is particularly useful to an executive who is likely to oscillate or vacillate in his choice of different strategies in a specific uncertainty situation. From the pay-off the possible strategies, the loss of regret, incurred in average expenses or profits by selecting a particular strategy instead of others under the same background conditions, are calculated and arranged in the form of a "regret matrix" as in Exhibit 13.2. The decision rule in such a case is to identify the maximum regret for each strategy and choose that strategy for which the maximum regret is least. In the example maximum regret for X is 87 and for Y is Rs 213. Choice of apparatus X minimises the maximum regret and apparatus X is to be preferred under these conditions.

CONCLUSION

These principles help the materials manager to minimise the ignorance associated with any purchasing decision-making problem under conditions of uncertainty. The executive process of decision-making under conditions of uncertainty could be summed up in the following diagram:

EXHIBIT 13.3

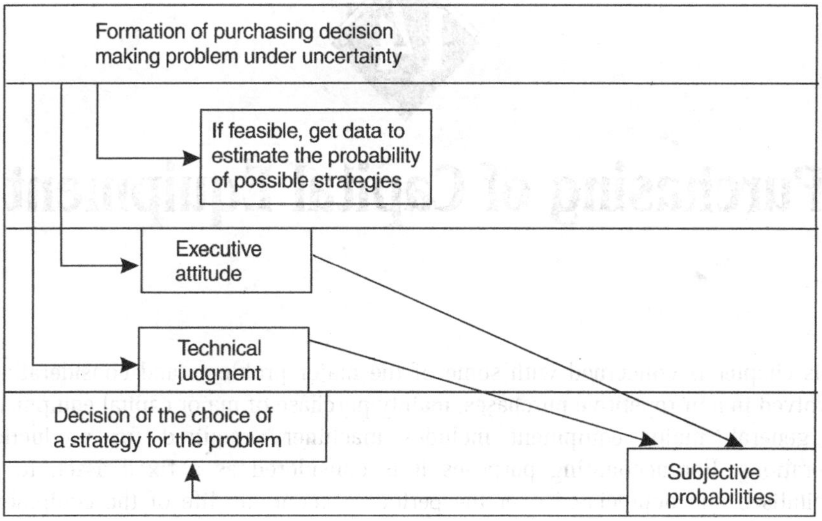

It would be interesting for an executive to try and apply these principles to some specific situation of his past or present experience. However, it should be mentioned that whatever methods are adopted in the process of decision-making, the executive must see that decisions are consistent with the following three broad principles on concrete quantitative data and are always directed towards action: (1) furtherance or continuation of the objective and the activities of the organisation; (2) maximum satisfaction to the largest number of persons involved in the organisation; and (3) demands of the environment in which the organisation operates.

14

Purchasing of Capital Equipment

This chapter is concerned with some of the major problems and considerations involved in non-repetitive purchases, mainly purchase of major capital equipment. In general, major equipment includes machinery required for production operations. For accounting purposes it is considered as a fixed asset, to be capitalised and depreciated over the period of economic life of the equipment. The purchase of capital equipment is usually differentiated both in procedures and policies from the purchase of consumption materials mainly because of the nature of large investment over a long duration.

SIGNIFICANT DIFFERENCES

The purchase of capital equipment usually occurs once in ten or fifteen years. The significant differences in the purchase of major equipment from the purchases of consumption materials are now briefly discussed.

1. Extent of Negotiations

Purchase of major equipment is a result of careful negotiations and analysis of various alternatives available to an organisation. The number of contacts and discussions with the vendors will be much greater compared to those for the purchase of consumption materials.

2. Lead-time Requirements

A unique feature in the purchase of capital equipment is the lead time required for its supply. Though very few equipments are standardised, most of the major equipments are custom built and require sufficient lead time for the suppliers to fabricate or manufacture and deliver.

3. Size of the Expenditure

This is the major consideration as the organisation would have to invest large amounts over a long duration. Moreover, the equipment directly influences the manufacturing operations of the organisation. It is natural that the purchasing of major equipment is a matter of significant concern to the top management.

4. Availability of Sources

Since many of the requirements of major equipment are specific and typical for different types of industries, the number of suppliers available is also very few and the organisation has to decide carefully a number of factors.

CONSIDERATIONS IN EVALUATION OF BIDS

As the purchase of capital equipment involves huge expenditure, the evaluation of bids of suppliers should be done in an extensive manner based on the following considerations:

1. Operating Characteristics and the Engineering Features

This is the most influential factor in selecting the supplier of major capital equipment. In almost all organisations, the initiation of capital equipment purchasing comes from the production department or the user department. Once the production or the user department establishes the need for the equipment and defines its functions, then the design and operating characteristics form the most important factors in the selection of the specific machine to be purchased. The number of manufacturers who are prepared to produce a piece of equipment that meets the requirements, both in design and operating characteristics, are very few and more so in the case of specialised equipment that has to be fabricated for one particular industry. The engineering features of the equipment should be compatible with the existing process and the specific question that should always be asked is: "Does the equipment fit in with the existing operations of the organisation and what are the modifications required in adapting the equipment to the existing system?"

2. Economic Analysis of the Investment

Economic analysis of the investment on purchase of a major capital equipment is to be done only after several acceptable pieces of equipment are found. One of the most important considerations is not simply to compare the cost of the equipment that is involved at the time of purchase, but to consider also the cost of modifications or additional fittings required, the total operating cost involved in the total plan of the equipment's economic life, and an estimation of cost of maintenance. Usually, cost of maintenance is to be included along with the cost of operation. The specific mention of the cost of maintenance is due to the fact that costs of operation can be forecast definitely while forecasting maintenance cost is difficult and subjective. Hence, in the economic analysis the total cost that is involved over the whole life of the equipment and the return it brings in the form of revenue or profit over its life-span, are considered.

Several methods of economic analysis are available for evaluating investment. Among these the most important are: (a) Pay-back period approach; (b) Return on investment; (c) Internal rate of return; and (d) Discounted cash flow method.

(a) *Pay-back period approach.* This approach is the simplest and most widely used method for evaluation of investment. It is the length of time required for the cash inflow resulting from an investment to equal the original outlay on the investment, that is, the time required to get back the initial investment on the equipment. This is mainly used in the case of purchase of equipments that have a short economic life and the organisation is of the view that the equipment should pay for itself. Hence, the shorter the pay-back period the better is the investment. The formula for pay-back can be given as: $t = C/I$ where t is the pay-back period in number of years, C is the initial capital outlay of the investment, and I is the average annual cash inflow expected from the investment. It is important to note that in the estimation of cash inflow, I depreciation should be included, as depreciation by itself is a measure of capital recovery.

In spite of its simplicity and wide use, this method suffers from a number of limitations. The first limitation is that it assumes uniform cash inflow each year and it cannot handle varying annual cash inflows. The second and most important limitation is that it ignores profits earned after the recovery of the initial investment. Hence, in the use of this method the tendency is always to choose investments which result in higher cash inflows in the beginning rather than those which give less inflows in the beginning and more later.

(b) *Return on investment approach (ROI).* Return on Investment (ROI) is just a reciprocal of the pay-back period, that is, ROI = I/C where I is the average cash inflow after taxes, and C is the original cash outlay.

Generally, there are two ways of estimating the return on investment: (i) Return on original investment; and (ii) Return on average investment.

In the first, the return is calculated on the actual value of the original investment, whereas, in the second method, the average investment, which is the mean of original investment and the salvage value expected at the end of economic life is taken instead of C. In both the methods, the average annual earning is taken to estimate the return on investment. This method again suffers from the limitation that it gives only the approximate rate of return and does not take into account the differences in the times at which investments generate their revenue.

(c) *Internal rate of return.* This method is understood under different names, discounted rate of return, yield of an investment, etc. It can also be defined as the procedure to find a rate of interest that will make the present value of the cash proceeds expected from an investment equal to the present value of the investment. It is the rate of discount which when applied to the future cash flows will equate their sum to the original price of the investment. It is usually a trial and error method. Probably, the method can be easily understood by the formula which is given as:

$$C = \frac{R_1}{1+r} + \frac{R_2}{(1+r)^2} + \frac{R_3}{(1+r)^3} + \ldots + \frac{R_n}{(1+r)^n} + \frac{S}{(1+r)^n}$$

where C is the initial capital outlay on the investment, r is the discounted rate of return, R_1, R_2 and $R_3 \ldots R_n$ are the cash inflows (or net savings) over the years 1, 2, 3, . . ., n (n being the economic life of the investment), and S is the salvage value of the investment (equipment) at the end of n years—the economic life. Thus, this method takes care of varying cash inflows and also the salvage value at the end of economic life.

(d) *Discounted cash flow method.* As against the above methods, the Discounted Cash Flow (DCF) method gives a true picture of the return on the investment as it takes care of the differences in times at which revenues are generated from an investment. The DCF method also uses the 'Present Value' concept in the evaluation of investments. All future revenues resulting from the investment are discounted so as to find the total present value of all these future earnings and compared with the value of the investment. If the total present value of the future revenues is more than the present investment then it is worth going in for the investment. 'Present Value', in simple terms, means "a rupee earned today is worth more than a rupee earned years later". Similarly, a rupee earned a year later is the same as Re 0.909 earned today. The present values can be explained directly through Exhibits 14.1 and 14.2. Exhibit 14.1 indicates the present value of Re 1 received in 'n' years, hence at different rates of discounting. For example, at a discount rate of 20 per cent, a rupee earned after one year is worth Re 0.833 earned today. Similarly, a rupee earned after 10 years is same as Re 0.162 only, if earned today. Exhibit 14.2 indicates the present value of Re 1 earned annually for 'n' years. For example, if on any investment we receive annual income of Re 1 for 10 years, then at a discount rate of 20 per cent, the present value of all these revenues over the 10 years is Rs 4.192. (The reader must have instantly found out that Exhibit 14.2 gives cumulative values over the 'n' years of values in Exhibit 14.1.)

In the Discounted Cash Flow Method, trouble may arise on the discount rate that is to be chosen. It is normal practice to choose the "cost of capital" or "opportunity cost" as the rate of discount for this method.

All the above four methods are explained better in an example:

EXAMPLE. A firm would like to buy a new milling machine costing Rs 1,50,000 with an economic life of 15 years. The milling machine is to be used for production operations which expect to bring up the firm's profits after tax as given in Exhibit 14.3 over the economic life of the equipment.

EXHIBIT 14.1
PRESENT VALUE OF RE 1

Years hence	1%	2%	4%	6%	8%	10%	12%	14%	15%	16%	18%	20%
1	0.990	0.980	0.962	0.943	0.926	0.909	0.893	0.877	0.870	0.862	0.847	0.833
2	0.980	0.961	0.925	0.890	0.857	0.826	0.797	0.769	0.756	0.743	0.718	0.694
3	0.971	0.942	0.889	0.840	0.794	0.751	0.712	0.675	0.658	0.641	0.609	0.579
4	0.961	0.924	0.855	0.792	0.735	0.683	0.636	0.592	0.572	0.552	0.516	0.482
5	0.951	0.906	0.822	0.747	0.681	0.621	0.567	0.519	0.497	0.476	0.437	0.402
6	0.942	0.888	0.790	0.705	0.630	0.564	0.507	0.456	0.432	0.410	0.370	0.335
7	0.933	0.871	0.760	0.665	0.583	0.513	0.452	0.400	0.376	0.354	0.314	0.279
8	0.923	0.853	0.731	0.627	0.540	0.467	0.404	0.351	0.327	0.305	0.266	0.233
9	0.914	0.837	0.703	0.592	0.500	0.424	0.361	0.308	0.284	0.263	0.225	0.194
10	0.905	0.820	0.676	0.558	0.463	0.386	0.322	0.270	0.247	0.227	0.191	0.162
11	0.896	0.804	0.650	0.527	0.429	0.350	0.287	0.237	0.215	0.195	0.162	0.135
12	0.887	0.788	0.625	0.497	0.397	0.319	0.257	0.208	0.187	0.168	0.137	0.112
13	0.879	0.773	0.601	0.469	0.368	0.290	0.229	0.182	0.163	0.145	0.116	0.093
14	0.870	0.758	0.577	0.442	0.340	0.263	0.205	0.160	0.141	0.125	0.099	0.078
15	0.861	0.743	0.555	0.417	0.315	0.239	0.183	0.140	0.123	0.108	0.084	0.065
16	0.853	0.728	0.534	0.394	0.292	0.218	0.163	0.123	0.107	0.093	0.071	0.054
17	0.844	0.714	0.513	0.371	0.270	0.198	0.146	0.108	0.093	0.080	0.060	0.045
18	0.836	0.700	0.494	0.350	0.250	0.180	0.130	0.095	0.081	0.060	0.051	0.038
19	0.828	0.686	0.475	0.331	0.232	0.164	0.116	0.083	0.070	0.060	0.043	0.031
20	0.820	0.673	0.456	0.312	0.215	0.149	0.104	0.073	0.061	0.051	0.037	0.026

EXHIBIT 14.2

PRESENT VALUE OF RE 1 RECEIVED ANNUALLY FOR *n* YEARS

Years hence	1%	2%	4%	6%	8%	10%	12%	14%	15%	16%	18%	20%
1	0.990	0.980	0.962	0.943	0.926	0.909	0.893	0.877	0.870	0.862	0.846	0.833
2	1.970	1.942	1.886	1.833	1.783	1.736	1.690	1.647	1.626	1.605	1.566	1.528
3	2.941	2.884	2.775	2.673	2.577	2.487	2.402	2.322	2.283	2.246	2.174	2.106
4	3.902	3.808	3.630	3.465	3.312	3.170	3.037	3.194	2.855	2.798	2.690	2.589
5	4.853	4.713	4.452	4.212	3.993	3.791	3.605	3.433	3.352	3.274	3.127	2.991
6	5.795	5.601	5.242	4.917	4.623	4.355	4.111	3.889	3.784	3.685	3.498	3.326
7	6.728	6.472	6.002	5.582	5.206	4.868	4.564	4.288	4.160	4.039	3.812	3.605
8	7.652	7.325	6.733	6.210	5.747	5.335	4.968	4.639	4.487	4.344	4.078	3.837
9	8.566	8.162	7.435	6.802	6.247	5.759	5.328	4.946	4.772	4.607	4.303	4.031
10	9.471	8.983	8.111	7.360	6.710	6.145	5.650	5.216	5.019	4.833	4.494	4.192
11	10.368	9.787	8.760	7.887	7.139	6.495	5.988	5.453	5.234	5.029	4.656	4.327
12	11.255	10.575	9.385	8.384	7.536	6.814	6.194	5.660	5.421	5.197	4.793	4.439
13	12.134	11.343	9.986	8.853	7.904	7.103	6.424	5.842	5.583	5.342	4.910	4.533
14	13.004	12.106	10.563	9.295	8.244	7.367	6.628	6.002	5.724	5.468	5.008	4.611
15	13.865	12.849	11.118	9.712	8.559	7.606	6.811	6.142	5.847	5.575	5.092	4.675
16	14.718	13.578	11.652	10.106	8.851	7.824	6.974	6.265	5.954	5.669	5.162	4.730
17	15.562	14.292	12.166	10.477	9.122	8.022	7.120	6.373	6.047	5.749	5.222	4.775
18	16.398	14.992	12.659	10.828	9.372	8.201	7.250	6.467	6.128	5.818	5.273	4.812
19	17.226	15.678	13.134	11.158	9.604	8.365	7.366	6.550	6.198	5.877	5.316	4.844
20	18.046	16.351	13.590	11.470	9.818	8.514	7.469	6.623	6.259	5.929	5.353	4.870

EXHIBIT 14.3

Year	*Profit after tax and depreciation* (Rs)
1	30,000
2	30,000
3	30,000
4	28,000
5	26,000
6	24,000
7	22,000
8	18,000
9	18,000
10	16,000
11	14,000
12	14,000
13	12,000
14	10,000
15	8,000
	3,00,000

Salvage value at the end of 15 years is expected to be Rs 5,000. The firm has alternate investment opportunities which can bring a return of 10 per cent on the investment. Should the firm buy the milling machine?

Solution

Total earnings Rs 3,00,000
Average annual earnings Rs 20,000

(i) Pay-back period $\dfrac{\text{Rs } 1{,}50{,}000}{\text{Rs } 20{,}000} = 7.5$ years

(ii) Return on investment:

(a) Return on original investment $= \dfrac{20{,}000}{1{,}50{,}000} \times 100$

$= 13.3$ per cent

(b) Return on average investment $= \dfrac{20{,}000}{77{,}500} \times 100$

$= 25.8$ per cent

(iii) Internal rate of return:

(a) On the basis of average income of Rs 20,000

$$1{,}50{,}000 = \frac{20{,}000}{1+\frac{r}{100}} + \frac{20{,}000}{\left(1+\frac{r}{100}\right)^2} + \frac{20{,}000}{\left(1+\frac{r}{100}\right)^3}$$

$$+\ldots+\frac{20{,}000}{\left(1+\frac{r}{100}\right)^{15}} + \frac{5{,}000}{\left(1+\frac{r}{100}\right)^{15}}$$

To determine r we will have to use a trial and error method.

If r = 10 per cent, the right-hand side can be calculated by using Exhibits 14.1 and 14.2 given earlier:

The right-hand side will be equal to:

Rs 20,000 received annually for 15 years + Rs 5,000 received after 15 years

= Rs 20,000 × 7.606 (from Exhibit 14.2) + Rs 5,000 × 0.239 from Exhibit 14.1

= Rs 1,52,120 + Rs 1,195

= Rs 1,53,315

This is slightly greater than Rs 1,50,000.

Hence the internal rate of return is slightly lower than 10 per cent.

(b) On the basis of varying incomes:

$$1{,}50{,}000 = \frac{30{,}000}{1+\frac{r}{100}} + \frac{30{,}000}{\left(1+\frac{r}{100}\right)^2} + \frac{30{,}000}{\left(1+\frac{r}{100}\right)^3} + \frac{28{,}000}{\left(1+\frac{r}{100}\right)^4}$$

$$+\ldots+\frac{8{,}000}{\left(1+\frac{r}{100}\right)^{15}} + \frac{5{,}000}{\left(1+\frac{r}{100}\right)^{15}}$$

For this, one would have to use Exhibit 14.2 again; by trial and error, the value of r is to be determined. If r is 10 per cent (the opportunity cost), then go in for the investment

(iv) Discounted cash flow method:

(a) On the basis of average incomes: present value of Rs 20,000 received annually for 15 years + present value of Rs 5,000 (salvage) value received at the end of 15 years

= Rs 20,000 × 7.606 + 5,000 × 0.239

= Rs 1,52,120 + Rs 1,195

= Rs 1,53,315.

Therefore, present value of future earnings – the initial investment

= Rs 1,53,315 – Rs 1,50,000 = Rs 3,315

Hence go in for the purchase of the equipment.

(b) On the basis of varying annual incomes:

EXHIBIT 14.4

Years hence	*Amount received* (Rs)	*P.V. factor from Exhibit* 14.1 *at* 10%	*Present value of the income* (Rs)
1	30,000	0.909	27,270
2	30,000	0.826	24,780
3	30,000	0.751	22,530
4	28,000	0.683	19,324
5	26,000	0.621	16,186
6	24,000	0.564	13,536
7	22,000	0.513	11,286
8	18,000	0.467	8,406
9	18,000	0.424	7,632
10	16,000	0.386	6,176
11	14,000	0.350	4,900
12	14,000	0.319	4,466
13	12,000	0.290	3,480
14	10,000	0.263	2,630
15	8,000	0.239	1,912
16	5,000 (salvage value)	0.239	1,195
	Total present value:		Rs 1,75,709

Therefore,

Present value of future revenues – Present investment

= Rs 1,75,709 – Rs 1,50,000

= Rs 25,709

Hence, go in for the purchase of milling machine.

Thus, anyone would have obviously noticed that the average income concept would in most cases give a wrong picture of the investment evaluation.

Apart from the above four methods, there are other methods for evaluating investments. Even in the case of equipment replacement problems, the above methods can be used by using 'net savings from the new equipment'. One of the most important methods in equipment replacement problems is the MAPI formula, prepared and published by the *Machine and Allied Products Institute.** An advantage of this method is that it compares the new equipment with the existing equipment to be replaced and determines the exact time when the replacement of

*The interested reader is referred to *Business Investment Policy* by G. Terborgh, Machinery and Allied Products Institute, Washington DC, 1958.

machinery is to be done. A detailed discussion of this method is beyond our scope.

3. Qualitative Considerations

Apart from the above two considerations, namely, operating characteristics and economic analysis of the investment, certain qualitative factors have to be considered before finalising the choice of the supplier for the capital equipment. The various qualitative factors that have to be considered are as follows: Reliability of the supplier in producing quality products and standing behind his guarantees, financial soundness of the suppliers and operations, and services rendered by the suppliers such as demonstration service, installation and commissioning, after-sales service, and replacement of parts.

In the purchase of capital equipment the qualitative factors play a secondary role and should be considered only after exhaustive analysis, based on operating characteristics and economic analysis.

PURCHASE OF USED EQUIPMENT

As a fundamental principle, any industry would prefer new equipment to used equipment. Used equipment, if available, is purchased in preference to new equipment, mainly because of economic considerations. The buyer under no circumstances is restricted to the purchase of new equipment only.

SOURCES OF USED EQUIPMENT

In general, the sources of used equipment are: (i) directly from the owner, (ii) used equipment dealers, and (iii) brokers and auctioneers. The majority of the purchase of used equipment is done from used equipment dealers and it is essential that every purchase executive in an organisation develops a list of sources of used equipment. The first source of used equipment mentioned above—the original owner—is usually preferred by most prospective buyers as they feel that it is advantageous to buy the machine from the original owner as they can see the machine in operation at the source and also learn details about the usage history of the machine.

In most cases used equipments do not carry any warranty from the supplier. But, in the case where the purchase is from used equipment dealers, dealer guarantee can be obtained for the equipment purchased from them. Generally, there are two types of used equipments available from used equipment dealers—(i) re-conditioned equipment, and (ii) rebuilt equipment. But these types of equipment, in general, carry dealer's guarantee for a special period. Brokers and auctioneers represent the most risky source as they do not carry any warranty. The other possible disadvantage is that used equipment purchased through brokers and auctioneers cannot be tested before.

Another obvious reason for the purchase of used equipment is that, in general, it adequately satisfies the buyer's needs. It is probably also easily adaptable to the existing operations. Another reason could be the financial position of the company which does not allow huge expenditure on the purchase of new equipment.

DRAWBACKS IN BUYING USED EQUIPMENT

One of the striking disadvantages of used equipment is the inability of a buyer to determine the true condition of the machine and to estimate the length of economic life of the equipment from the time of purchase onwards. It is very essential that production, design and maintenance department personnel are involved in the purchase of used equipment so that a rough estimate of the true condition of the used machine can be made. The buyer should try to obtain the relevant equipment records and the history cards of the equipment so as to analyse the performance of equipment over its life and forecast its future performance. Another drawback is that the suppliers do not provide service for used equipment after the sale. In addition, all transportation, handling, installation and start of cost as well as risks of breakdown are to be borne by the buyer.

PURCHASE VERSUS LEASE

The question of leasing equipment arises because of: (i) availability of funds, (ii) the duration for which the equipment is required, (iii) the necessity of the organisation to obtain the equipment immediately (whereas purchase of new equipment may involve a lot of procedural and policy delay), and (iv) capital equipment is available only on lease.

The major considerations of leasing as against purchasing capital equipment are:

1. It may be very economical to lease the equipment when it is needed for a relatively short period. On the other hand, since most leases are for definite terms and terminable on short notice, it may result in depriving the lessee of a needed facility.
2. In the context of fast changing technology and obsolescence of equipment, leasing could avoid the purchase of a less efficient piece of equipment.
3. In general, in a lease the lessee bears the dismantling, loading and unloading charges and transportation expenses both ways, from lessor's location to lessee's and return, whereas in the purchase of new equipment, the equipment is generally sold at a delivered price.
4. The rent of leased equipment is an item of expense for tax purposes while the purchase price is an investment to be depreciated over a number of years. That is, as long as lease payments exceed the value of allowable depreciation, and additional tax benefit is accrued.
5. One of the most important disadvantages of leasing is that the lessor retains control of the equipment and not the lessee. This loss of ownership places certain restrictions in the manner in which the equipment is to be operated and it is obligatory on the part of the lessee that he allows the lessor access to the equipment for inspection at any point of time.

ROLE OF THE PURCHASING MANAGER

Unlike production materials, purchase of capital equipment is decided by a top management committee. The committee usually consists of members representing Production, Design and Engineering, Finance and Purchase. The recommendations of this committee are then finally approved by the General Manager or the Chief Executive of the company before the capital equipment is purchased.

The next important role of a purchase manager is to supply the qualitative information regarding the potential suppliers. He is in the best position to gather information regarding the image, the technical capabilities, the financial soundness and the reliability of the supplier. He should also find out whether the supplier is capable of providing technical know-how mainly in regard to the installation, commissioning and after-sales service. It is also his responsibility to check a supplier's *locus standi.* A purchase manager has to get complete details regarding offers of potential suppliers, tabulate them and supply them to the finance department so that they can make a thorough analysis. After the top management decides on a supplier, the purchase manager assumes the customary responsibility of negotiating the final price, terms and conditions of delivery, after-sales service and supply of spare parts. He should, also, in collaboration with the finance department, come up with suitable payment terms with the supplier. On the whole, the role of a purchase manager in the procurement of capital equipment is that of a liaison man.

15

International Purchasing

NEED

For a growing economy like India's, achieving self-sufficiency should be the goal but not at the cost of imports. Importing is an essential economic function which cannot be completely eliminated. Ricardo's principle of comparative advantage states that it would be beneficial for an economy to concentrate on the production of items in which it specialises, export these items and import its requirement of other items. The principle, though not totally practical, cannot be dismissed. International trade is undoubtedly governed by political motives and, as such, a country cannot totally rely on another country for its requirements of specific items. It would be equally fruitless to undertake import substitution at any cost. The Government of India from time to time reviews the import policies considering various aspects such as trade balance, raw materials position, etc. Recently the import policy has been liberalised, as shortage of raw materials has led to under-utilisation of capacity. In some cases, automatic licences will be issued to cut down procedural delays. Import policies as such are not static and changes in policies and procedures are very important to industries which depend on imports for their production.

PROCEDURE

If a company is importing for the first time, the first and foremost step for it would be to check whether the item is allowed to be imported. This information will be available in the Red (Blue) book. If the item is banned for imports (on the ground that it is indigenously available), the starting point is to get a list of all local manufacturers and ask for quotations. If the manufacturers are not in a position to supply, then a written statement to that effect should be obtained from them. These statements should be presented to the Government to get the clearance for imports. The government, on its part, might give names of some more possible suppliers or could write to the suppliers themselves and then possibly clear the import of the items. Companies can get import entitlements on the exports even if they be indirect.

In case the import clearance is obtained or the item is not banned for imports, a company has to go through the following stages to accomplish the purchase:

1. ***Locating the foreign source of supply.*** This has to be done by contacting the Ministry of Trade, Foreign Consulates and Embassies, Indian Consulates in foreign countries, etc. A foreign consulate or embassy has a commercial attaché who is well versed with the list of suppliers in his country.

2. ***Procurement of the item.*** At this stage the importer will be involved with foreign source of supply, manufacturing unit of his company, engineering department and governmental institutions like DGTD, MMTC and STC.

3. ***Documentation.*** Special documents, not used for indigenous purchases, are required. The commonly used documents are: (a) Bill of lading, (b) Invoices, (c) Certificate of origin, (d) Weight certificates, (e) Insurance policy, (f) Markings of packages.

NATURE OF DOCUMENTS

Bill of Lading

The *bill of lading* is the most important document which accompanies bills of exchange drawn under letters of credit. It is evidence of the fact that goods have been despatched by the exporter and gives the importer title to the goods and enables him to claim them on arrival at the destination.

A bill of lading is a document signed by the master of a ship or by the shipowners or their agents acknowledging the receipt of certain specified goods for carriage and embodying an undertaking that the goods will be delivered to the consignee, or to his order or assignee, or merely to order. The bill specifies the port of shipment, the destination and the condition under which the goods are received for carriage. It is at the same time: (i) a receipt for goods delivered to carrier for transportation, (ii) a contract between the shipper and the carrier for transportation of the goods and then delivery to the consignee or his order, and (iii) a document of title to the goods, giving the holder title to the goods mentioned.

Bills of lading are prepared by the shipper on forms supplied by the shipping company. Normally, a bill of lading shows the date and place of shipment, the name of the carrying vessel, the name of the consigner and the consignee, the post of destination, the number, contents, and identification mark of the goods shipped, and the amount of freight 'paid' or to pay. The detailed particulars of the number of packages, the mark they bear, their contents and the amount of the freight on them are entered in the space provided for them.

Bills of lading are usually made out in sets of four, three copies signed and one left unsigned; the last being retained by the ship's master for his own use. In the commercial contracts, a set usually means a set of three signed copies, though there may be more or fewer copies according to the requirement laid down in a particular contract. Goods are released at destination only against surrender of the bill, and when one copy has been surrendered the other copies become void.

The bill of lading possesses some characteristic of a negotiable instrument. It can be transferred by endorsement to another person. Goods shipped under a bill of lading are often deliverable to a person named or to his order, or just to order, in which case the title to the goods can be transferred by endorsement and delivery of the bill of lading.

In the case of a bill which does not name any particular consignee, transference can be by mere delivery. It is, however, not a negotiable instrument in exactly the same sense as a bill of exchange. When it is transferred, the transferee acquires no better title than the transferer. In the case of a wholly negotiable instrument the transferee gets a title without any defects of the transferer's title. A bill of lading is transferred "subject to equities", i.e. even if the transferee takes it for value and in good faith, he will not obtain goods if the instrument is in any way affected by fraud. The bill of lading is, therefore, only a quasi-negotiable instrument.

If a foreign trade contract provides for payment of freight by the consigner, the bill of lading must be marked "freight paid" or freight receipt must be attached. If words such as "freight prepayable" or "freight to be prepaid" or words to similar effect appear on the bill, they will not be accepted as evidence of the payment of the freight. The carrier has a lien on the goods which he carries, and he has the right to refuse release of goods until his charges are paid. Therefore, whosoever comes into possession of a bill of lading for consideration acquires the document subject to the carrier's lien representing his charges.

Clean and foul bills of lading. While it is not a carrier's responsibility to examine and verify the contents of the packages consigned for shipment, he does see that the packing is good enough for the requirements of the contents and to stand the handling the goods will receive before they are delivered. A bill of lading which indicates that the goods are in apparent good condition without any qualification is known as a **clean** bill of lading. If, on the other hand, the bill bears some remark relating to a defect in packing, such as 'three bags torn', 'some cases broken', 'received in damped condition', 'goods not sufficiently packed', etc., it is known as a **foul** or **dirty bill of lading**. A 'foul' bill of lading is not an acceptable document, unless specifically provided for in the foreign trade contract. Such remarks protect the carrier against any claim for losses attributable to bad packing. They also protect everyone else who comes in possession of the bill of lading for value. Moreover, an insurance company also does not provide any cover for the part of the cargo which is poorly packed.

Invoices and certificates. Among the documents that accompany bills of exchange drawn under letters of credit in foreign trade transactions are the commercial invoice, the consular invoice, the certificate of origin and the packing list. There are in effect detailed descriptions of the goods shipped, their prices and other charges, the country of origin of the goods, and other particulars of the transaction. Which of these documents are included and in what form is a matter of contract between the buyer and the seller.

Commercial invoice. The Commercial Invoice describes the merchandise, indicates the price and other details of the transaction. The name and address of the buyer and the seller, the vessel of shipment, port of discharge, shipment, export and import permit numbers, the number of contract and invoice and the other essential details of the transaction are recorded on the invoice. The financial terms of the sale, such as whether under letter of credit, sight drafts, etc., are given. The number of packages and the identifying marks and numbers on them are clearly indicated. The goods contained in the packages are detailed and prices are given item by item, the trade discount is deducted from the gross aggregate price of the goods, and a total net value is shown. Any other charges to be included under the contract such as for packing, freight, insurance, the cost of preparing a consular invoice, etc., are added to the total.

The commercial invoice is not a document of title, nor is it negotiable. Its main utility is as a book-keeping device for both seller and buyer. It is a double check on accuracy, a check of actual goods against the invoice and the invoice against the copy of the order. It also indicates whether the agreed price has been charged, and enables costs of any damage to merchandise to be calculated. Banks check the amount of the invoice against the letter of credit, if a letter of credit is used in financing the transaction. Normally, the amount of the invoice must be within the limit set by the letter of credit, and any excess may indicate inaccuracy in the invoice, a misunderstanding or deliberate overcharging the invoice by the seller. The banker then warns the buyer accordingly and the latter can take steps to get the discrepancy rectified.

Certificate of origin. The laws of some countries require a certificate of origin of imported goods to be produced before clearance of goods by customs and assessment of duty. An importer, therefore, may request the exporter to send a certificate of this kind along with other documents. The certificate is usually required where goods from certain countries receive preferential treatment or the import of goods from some countries is partially or wholly prohibited. The country where the goods are produced should, therefore, be known to the customs authorities. Sometimes, the certificate of origin is endorsed on the back of the relative invoice, in which case the whole document is known as a "Certified Invoice".

Packing list. A packing list serves to indicate the exact nature, quantity, and quality of the contents of each package in a shipment. The list helps the importer to identify the goods and check them against his own order. Banks may require such a list when they have financial interest in the merchandise. Clearance of goods through customs is also facilitated by packing lists.

Nature of insurance policy. Marine Insurance Policy is a contract between the insurer and the insured whereby the former, in consideration of the payment of a premium by the latter, agrees to indemnify the latter against loss incurred by him in respect of goods exposed to "perils of the sea" or to the particular perils insured against. For a policy to be valid, the insured must have an insurable interest in the goods subject to insurance, i.e. he benefits from the existence of the goods and suffers a loss if the goods are damaged or lost. The insured need not have this interest at the time of insuring, but must have acquired it at the time of putting a claim under the policy for damage or loss of the goods.

The insurer, known in marine insurance parlance as the underwriter, is the party providing insurance cover, and it may be an insurance company transacting marine business or some firm of underwriters. A proposal for marine insurance is made through a marine insurance broker, who writes the particulars of the proposed insurance on a ship, which he submits to some underwriter. If the latter accepts the risk he signs the slip, and the insurance is in force from that time though the actual policy is delivered later. Further details of reference may be made to the chapter on Insurance Buying.

Markings of packages. A packing list serves to indicate the exact nature, quantity, and quality of the contents of each package in a shipment. The list helps the importer to identify the goods and check them against his own order. Banks may require such a list when they have a financial interest in the merchandise. Clearance of goods through customs is also facilitated by packing lists. The packages should be suitably marked with address, dimensions, weight and other particulars. The packings should be sewn correctly and in conformity with instructions in the order.

Payments. The importer is involved with banks, Ministry of Finance, and the Reserve Bank. The documents involved are: (1) Letter of credit/drafts, (2) Terms of payment, (3) Rate of exchange, (4) Payments under various credits and trade agreements.

Letter of Credit

A letter of credit is an arrangement whereby the obligation to pay an exporter is undertaken by a bank. The bank's credit is available to an importer who is not known outside and who otherwise would not be trusted with goods by a trader in another country. If a letter of credit is received from a bank the credit standing of the importer is of little consequence. What becomes important to the exporter is the credit standing of the bank issuing the letter of credit.

The importer requests a bank in his country to issue a letter of credit in favour of the exporter. The bank cables or writes to a correspondent or some other bank in the exporter's country that a credit has been opened in favour of the exporter, and the latter bank informs the exporter about it. The advising bank informs the exporter that it will honour drafts drawn by the latter under terms and conditions laid down in the letter of credit. The liability for payment is really that of the bank issuing the letter of credit and the bank that honours the drafts drawn by the exporter, gets reimbursed by the issuing bank.

Essential functions of a letter of credit. 1. The name of the bank issuing the letter of credit is stated on the top of the instrument. The exporter receiving the letter should see whether the bank is acceptable to him or not.

2. The instrument mentions the name and full address of the beneficiary, i.e. the seller in whose favour it is issued.

3. The amount for which the credit is issued is clearly stated. This amount is sometimes calculated exactly on the basis of the quantity of merchandise to be

shipped. Banks are then not willing to accept and honour drafts for a smaller or larger amount. To avoid this difficulty some flexibility mentioned in the amount is introduced by using the word "about" with the sum.

4. The name of the importer for whose account the credit is opened is mentioned in the letter of credit. The credit standing of the importer is of little consequence to the seller if he is satisfied with the issuing bank. In fact the importer may not be known to the exporter at all.

5. Another important point to be clearly stated in the letter of credit is the tenor of the draft or drafts to be drawn under it. The draft may be a sight draft. This depends on the initial contract regarding the transaction between the exporter and the importer.

6. The letter of credit lists the documents to be attached to the draft drawn under it. The list is made out by the importer and given to the bank of issue. The documents are such as clearly indicate the discharge of contractual obligation by the exporter. Any discrepancy between the documents that accompany the draft and as those specified in the letters may result in a refusal by the advising bank to honour the draft.

The documents specified make clear the type of bill of lading, the conditions of marine insurance, if ocean freight is to be prepaid, and other details which do not have any misunderstanding in the mind of the exporter. It is possible that the documents demanded may not be in accordance with the custom and usage regarding them. In such a case, the exporter may object and should communicate with the advising bank. To speed up matters, direct communication with the importer is often resorted to.

7. The letter of credit has an expiry date, i.e. a date up to which the instrument is valid. A time or an exact date is mentioned by which the whole operation should be completed. The date given is normally one which will allow the exporter to conveniently ship the merchandise under the contract and to get all the required documents within the period. Again, if the time given is not enough for the exporter he may communicate with the issuing bank or the importer and get the date changed.

8. The type of credit opened, whether revocable or irrevocable, confirmed or unconfirmed, is stated on the instrument.

9. The letter of credit mentions any other conditions that may be demanded by the importer and that may have been settled between the importer and the exporter.

Transportation of product to end user. The buyer will be involved with the shipping company, forwarding agents, insurance company, custom and import authorities. As we have seen, the documents and insurance policy cover this area.

PROBLEMS

The Government policy being one of protecting scarce foreign exchange, the procedures for clearance of imports are naturally very rigorous. It takes about 6–9 months to get the clearance from the authorities. For example, in the case of public sector undertakings there are about 10 different import procedures depending on the type of item imported (capital equipment, spare parts and raw materials) and the type of industry (priority, under the purview of ministry of steel and mines).

The imported quantity has to be one year's requirement and the renewal of import licence is based on the previous year's consumption. The minimum average inventory a company has to hold is six months' requirements. However, in an attempt to play safe, companies carry heavy stock and one organisation is reported to have more than four years' requirement. The policy of conserving scarce foreign exchange has resulted in the locking up of the exchange in the form of inventory holdings.

TRADING ORGANISATIONS

To aid the import of certain basic raw materials like copper, zinc, tin, lead and nickel, the government has set up a trading organisation—Minerals and Metals Trading Corporation of India. It is possible for companies to arrange for the procurement of basic metals through MMTC. The delivery arrangement could be staggered and hence the investment in inventories can be brought down. For the imports of certain other materials, like caprolactum, chemicals, raw cashew nuts, etc., there is another government undertaking—the State Trading Corporation.

The above mentioned trading organisations require to undertake long-term contracts with international suppliers to safeguard against fluctuating conditions in supply and price. Both MMTC and STC are the important canalized agencies and account for about 50 per cent of the imported value.

With the strict import procedures and the heavy supply lead time, scarce foreign exchange is locked up in inventories. The government has realised the situation and is making attempts to cut down the time for clearance of imports to 3–4 months. It has also set up two trading organisations which provide help in the import of basic raw materials. The Government of India is taking several measures like automatic licences for speeding up the licensing procedures. The hope of import clearances being made faster and the measures taken for import substitution are healthy signs for a purchase executive.

International Credits

There are a large number of credits and loans available from different countries and also from the World Bank. The International Monetary Fund is helping the developing nations afflicted with the world oil crisis, through special drawing rights, CSDR facilities. The United Kingdom has provided loans for imports into India from the UK. Some of these are UK/India Capital Investment Loan, UK/India Mixed Project Loan, UK/India Mixed Imports Loan. Import licences are issued for import of maintenance spares as well as capital items under these loans. Procurement is limited to the UK. In some of the maintenance loans, a specified amount is reserved for British-oriented firms for import of maintenance goods. Similarly United States has also given several loans for importing goods in India. The developed countries have also formed Aid India Consortium to give loans and grants to India. Towards the end of 1973, the Government of India had entered into a 15-year economic pact with Russia and Czechoslovakia which envisaged substantial increase in trade between India and these nations. Assistance is also provided by the International Development Association. Under this credit, materials, components and spare parts required by certain specific manufacturing industries only are allowed to be imported. Procurement is permitted from any of the countries who are members of the World Bank.

There are other international organisations like the UNCTAD—United Nations Conference on Trade and Development, and the UNIDO—United Nations Industrial Development Organisation, which provide advisory services not only to developing countries but also to international organisations like the International Federation of Purchasing and International Organisation of Standardisation, etc. UNIDO, for example, has prepared a manual on Purchasing of International Equipment which will be of use to buyers who conduct international purchasing operations. If there is a question of dispute on tariff, GATT will assist the indenting party.

Many of these international organisations have their offices in Geneva, Switzerland, and the International Federation of Purchasing has consultative status to confer with these organisations on any problems which their member associations have. These organisations can be contacted for help through local Purchasing Associations.

Reference may be made to ISO—International Standards Organisations—by parties having any disputes or difference of opinion with regard to standards on specifications in international buying.

16

Import Substitution: Prospects and Retrospects

The policy of import substitution should be enshrined as a major national objective in a developing economy like India. As far as practicable the process of import substitution is the resultant effect of many national policy instruments aimed at achieving self-reliance as well as a favourable balance-of-payments position. Import substitution implies substituting what is imported by indigenous items as far as practicable. The detailed policies and programmes of various government departments like the Planning Commission, Department of Science and Technology, DGTD—Directorate General of Technical Development, CCIE—Chief Controller of Imports and Exports, MMTC—Metals and Minerals Trading Corporation, STC, etc., should be coordinated to achieve this laudable objective. The various policy instruments, enacted by Parliament, such as Industrial Policy Resolution, the Industry's Development and Regulations Act, the Five-Year Plans, Guidelines for Industries, Annual Import Trade Control Policy, etc., are supposed to be focussed towards the aforementioned national objective of self-reliance.

THE FACTS

It is rather unfortunate, however, that the balance-of-payments position has consistently worsened—obviously due to several factors—during the last three years. In the preceding period, i.e. during 1969–72, the average annual imports were of the order of Rs 1,650 crores with a deficit of Rs 110 crores. The picture in the last three years is given in Exhibit 16.1 (*Source*: Reserve Bank of India Bulletin, June 1975). The import value index—base 1958—has shot up to 333 points in 1974–75 from 226 points in 1973–74.

In view of the fact that the present trade deficit is around Rs 1,100 crores, the country has to take suitable pre-emptory measures to visualize an improved situation for the current year because of the 10 per cent increased petroleum price of 11.51 dollar per barrel announced by the OPEC—Organisation of Petroleum Exporting Countries—in September 1975. This increase alone would further deteriorate the deficit by 108 crores rupees per annum, i.e. 54 crores for the six months—October 1975 to March 1976. This energy crisis has pushed India into the "fourth world". This group comprises of the third world countries that the

EXHIBIT 16.1
BALANCE OF PAYMENTS

(*Rs crores*)

Year	1972–73	1973–74	1974–75	*Estimates of* 1975–76
Imports	1830	2920	4350	4600
Exports	1950	2520	3250	3600
Balance of trade	+ 130	– 400	– 1100	1000

EXHIBIT 16.1A
BALANCE OF PAYMENTS

Year	1979–80	1980–81	1981–82	1982–83	1983–84 *Estimates*
Imports	9022	12435	13680	14275	13500
Exports	6459	6709	8634	8808	8500
Balance of trade	– 2563	– 5726	– 5046	– 5467	– 5000

A pictorial presentation of the above is available in Chart 16.1.

CHART 16.1
BALANCE OF PAYMENTS 1972–76 (VALUE: RS, CRORES)

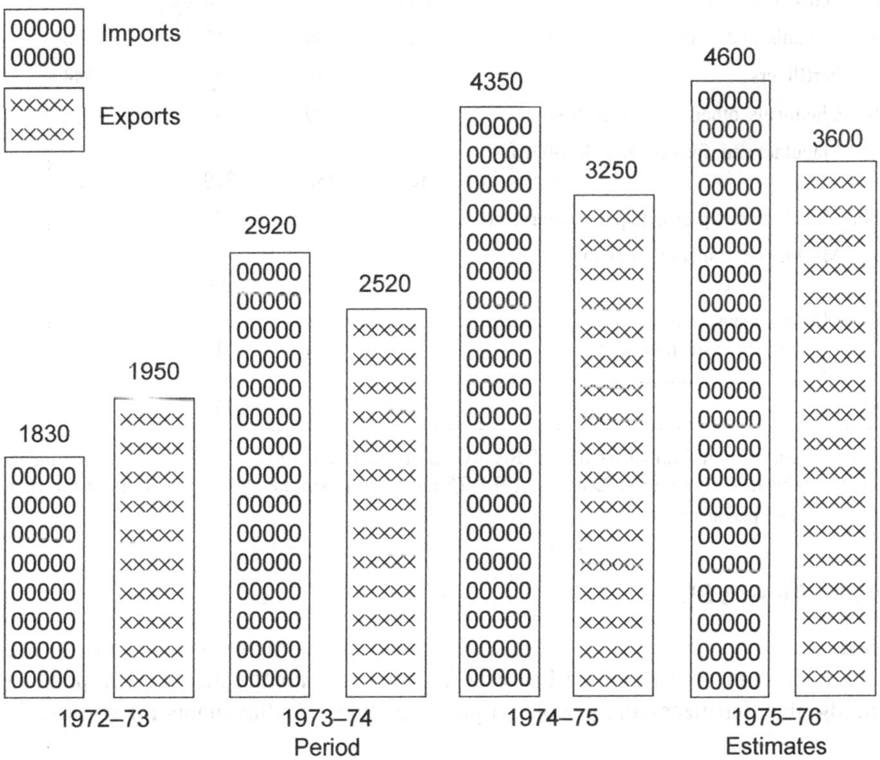

energy crisis has caused to further slip down economically, and experience great difficulty even in servicing foreign debts. Further, available information and the performance during April-September 1975, as well as our studies in Administrative Staff College of India, indicate that the imports for the year 1975–76 will be of the order of Rs 4,600 crores and the exports will be about Rs 3,600 crores, thereby resulting in the same deficit as existed during the previous year (we do observe that the exports in some sectors have recently risen by 15 per cent over the previous year).

The position with regard to import of other commodities in the current year, we feel, will maintain broadly the present trends—for the year 1974–75—as indicated in Exhibit 16.2.

EXHIBIT 16.2
IMPORTS: COMMODITY GROUPS

(*Rs, crores*)

Commodity groups	1972–73	1973–74	1974–75	*Per cent increases of* 1974–75 *over* 1973–74
1. Food and live animals, dry fruits and milk products	160	545	780	+ 43
2. Crude materials like wool, cotton, oilseeds, etc., except fuel	190	185	215	+ 19
3. Petroleum, lubricants and minerals	205	560	1,135	+ 103
4. Animals and vegetable oils and fats	25	65	45	– 30
5a. Fertilizers	90	160	390	+ 144
5b. Chemicals other than fertilizers	160	190	285	+ 50
6a. Manufactured ferrous and non-ferrous items	320	380	600	+ 58
6b. Manufactured glass, paper, pottery, etc.	115	150	165	+ 10
7. Machinery, transport and electrical equipments	515	630	690	+ 10
8. Miscellaneous manufactured items, scientific instruments and others	50	55	45	18
Overall:	1,830	2,920	4,350	+ 49

Note: Figures from Annual Report of Reserve Bank of India, July 1975. The figures rounded off to the nearest five crores. Data for 1974–75 have been extrapolated on the basis of the first ten months' performance.

THE ANALYSIS

It is observed from Exhibit 16.2 that 75 per cent of the increase in imports during 1974–75, compared to the earlier year, is accounted for the increased imports in foodgrains, fertilizers and petroleum products. Even if adjustments for worldwide inflationary phenomenon and price levels have been made, the volume of imports has remained constant and in some cases might have even increased.

Region-wise analysis of imports indicates that during the first half of 1974–75—April to December 1975, Indian imports from all regions increased. The rise in imports from Asia and Eastern Europe was highly significant, with the Asian region becoming India's foremost supplier, accounting for about 45 per cent of the country's import trade. Included in this region are the oil imports from Iran, Iraq and Saudi Arabia. Western Europe including EEC countries accounted for over 20 per cent of India's total import trade. The imports from Eastern Europe and USSR in the 6 months accounted for Rs 330 crores, whereas North America accounted for 15 per cent of India's total imports.

THE AGRICULTURAL SCENE

We would now like to classify the imports as (a) those arising due to inadequacy of natural resources, and (b) those due to rapid industrial growth. For instance the imports of petroleum products and non-ferrous metals fall in the former category. However, the increases in imports of fertilizer, ferrous metal and food are due to the stresses and strains of the economy and due to effects of faulty planning. Admittedly, imports of items such as food, petroleum and fertilizer should be considered inescapable imports in the short time horizon. Even though agriculture was given great prominence and rightly so, in the five-year plans, and the country gloated over the "green revolution" years ago in 1968, we still continue to import two to eight million tonnes of foodgrains year in and year out. In this context, the innumerable Agricultural Colleges and large number of Agricultural Universities should play a more positive and dynamic role towards better productivity in the farm sector. Further, it is worth observing that the annual budget of ICAR—Indian Council of Agricultural Research—is about Rs 35 crores. We really wonder when these scientists will be able to develop suitable strains that will withstand the vagaries of the monsoon and flourish in the Indian soil!

In order to arrest the massive increase in foreign exchange needed to finance food imports, it is essential that a suitable machinery be set up to procure the marketable surplus and build adequate buffer stocks for meeting the demands of a lean production season. Obviously, the farmer should be given remunerative prices so that he does not switch over to cash crops. It may be remarked here that the total foodgrain storage capacity, with Central Warehousing Corporation, State Warehousing Corporation and Food Corporation of India, has increased from six million tonnes to thirteen million tonnes. It is indeed heartening to note in this context that the Government of India envisages an additional five million hectares to be brought under irrigation. Further, the recent oil discoveries, both off-shore and on-shore, provide a silver lining in the dark clouds and are indeed very good auguries both in the petroleum and fertilizer sectors.

THE CANALIZED AGENCIES

The Governmental sector has to play a very crucial role in economising imports, as it accounts for about 75 per cent of the total values of import licences issued. This is more so with the case of canalized agencies such as MMTC and STC. It is estimated that imports through these two agencies in this year will total to about Rs 1,500 crores. It is too well known that these two agencies in the recent past had accumulated huge stocks of raw materials, essentially because the purchases were not carried out in a planned fashion. In view of the huge amount of imports involved, it is hoped that these trading agencies will develop adequate materials intelligence as well as suitable management information system covering aspects such as import substitution, world commodity prices, demand pattern supply position, etc., so that imports will be arranged in the most economical manner. This point has also been emphasised by the working group appointed by Government of India for building adequate stocks of key industrial raw materials. This group consisting of the Chief Economic Advisor as well as Senior Officials in the Centre has advised that the State trading agencies should avoid the practice of rushing to buy the materials at the most inappropriate time!

IMPORT SUBSTITUTION—THE INDUSTRIAL SCENE

We shall now discuss the prospects of import substitution in the industrial sector. Any developing nation trying to industrialise rapidly by attempting to leapfrog different stages of industrial growth, will have to import modern machinery as well as technical know-how. This process will undoubtedly result in a great strain on the scanty foreign exchange resources. However, it is heartening to note that India has already passed through the stage of importing complete production plants. In the industrial sector the imports are broadly restricted to spare parts, components, key industrial raw materials, special steels and basic chemicals. From Exhibit 16.2, it can be seen that the total value of imports during 1974–75 for chemicals, manufactured ferrous and non-ferrous items and machinery—categories 5b, 6a and 7—amount to Rs 1,575 crores. This, together with manufactured glass, paper, pottery and miscellaneous items, accounts for Rs 1,785 crores or about 40 per cent of total imports. Some of the above imports are through canalized agencies. Hence import substitution in the industrial sphere assumes a significant role.

We are really happy to note that India is now producing complete plants in the fields of textiles, cement, paper, with tapering foreign exchange content. In the coming years, India will be self-sufficient in making steel plants, power generating equipments, tyre-making machines, defence equipments and other core sectors of the economy. The 2000 MT press made by HMT for BHEL, Trichy, is a case in point.

In spite of laudable achievements of large public sector undertakings like BHEL—Bharat Heavy Electricals Limited, HMT—Hindustan Machine Tools, IPCL—Indian Petrochemicals Limited, IDPL—Indian Drugs and Pharmaceuticals Limited, BEL—Bharat Electronics Limited, etc., the foreign exchange content of these companies is fixed around a certain percentage—varying from ten to twenty-five per cent of value of production. Such figures for a public sector unit are given in Exhibit 16.3. This is because technology does not remain static. In order to keep pace with international developments, they cannot avoid imports. Industry can come out of this dilemma if it sets its objective as self-sufficiency in foreign exchange through exports.

EXHIBIT 16.3
PROGRESS OF IMPORT SUBSTITUTION IN A PUBLIC SECTOR ENTERPRISE

(Rs, lakhs)

Year	*Value of production*	*Foreign Exchange Content*	
		Value	*Percentage*
1956–57	2	0.5	25
1957–58	11	3	29
1958–59	68	21	31
1959–60	110	31	28
1960–61	131	57	44
1961–62	240	83	36
1962–63	311	109	35
1963–64	495	243	50
1964–65	521	172	33
1965–66	935	289	31
1966–67	1,016	305	30
1967–68	1,663	562	34
1968–69	1,954	694	35
1969–70	2,253	727	32
1970–71	2,806	805	29
1971–72	3,230	853	23
1972–73	3,733	903	24
1973–74	4,282	1,039	24

Note: The criteria for evaluating the achievements in import substitution should be based on decreasing value of the imported content *per unit* of production over period of time.

Strategy for Import Substitution

A new strategy for import substitution has to be based on (a) enlightened corporate policy of source development, (b) standardisation of equipments and designs, and (c) national technology policy.

Many organisations like BHEL, TELCO, HMT, L&T, the TVS group, etc., have established company policies for source development. Financial and technical assistance are provided by some of these companies for developing indigenous suppliers. Our experience and interaction with senior executives attending management development programmes in the Administrative Staff College indicate that many organisations pay only lip service to source development and they strongly feel that tangible results will be obtained only if the developments of suppliers are interwoven with the long-term corporate plans of the buying organisations.

Source Development

In developing indigenous sources the companies have to consider initial problems relating to high cost, poor workmanship, need for technical help, availability of raw material, etc. These problems in the gestation period will arise only when the supplier is building up his competence. Hence the companies should take a long-term view and should go out of their way to help the supplier technically and financially. A few companies even go to the extent of allowing the sophisticated imported machines to be opened up and the drawings scrutinised by the suppliers. Some manufacturing organisations encourage their own craftsmen and engineers to become entrepreneurs and suppliers. The concept of the industrial estate is a consequence of this logic. However, some of the industrial estates are not flourishing in the highly industrialised parts of the country like Bangalore, Poona, etc., because they are asked to concentrate only on minor jobs and sufficiently challenging orders are denied to them. In this context, we should keep in mind about the capabilities and achievements of the workshops in Jallandhar and Ulhasnagar.

Standardisation

Standardisation has been accepted as a basic necessity for mass production. It is a continuous and time-consuming process. In the case of industrially developing countries, a systematic approach for establishing priorities and a total view of industrial sectors is called for. Due consideration has to be given for the rapid advances taking place in technological fields in the world. While we are glad to note that the Indian Standards Institution has developed over 7000 standards, it has to go a long way in organising cells within I.S.I. for evolving standards in each industry. Perhaps this institution may have to reconsider its priorities and tackle the vast confusion of standards in the engineering industry rather than evolve standards for trousers and shirts! The Indian industries on their part should make a very concentrated effort to adopt Indian standards and replace foreign specifications. Industry groups like electronics, heavy engineering, light engineering, chemicals, etc., should take initiative in establishing their own working groups to evolve inter-company standards, instead of waiting for Indian Standards Institution to give them printed specifications on a golden platter! It is heartening to note that a group of public sector companies, in the electronics field like Bharat Electronics, Indian Telephone Industries, Electronics Corporation of India and Hindustan Aeronautics Limited, have already started work in this direction. These companies have put in some efforts for standardisation of components. It is also necessary to standardise the design of production machinery.

Large organisations having collaboration with more than one country need to have a close look at the production line with a view to standardisation. The above approach, being just initiated in BHEL and SAIL, should be followed by other large undertakings.

Role of Materials Managers

Materials managers who are in charge of source development and import substitution can play a significant role in this direction. They should encourage the engineers to adopt value analysis techniques which is a very fruitful area for import substitution. In this context the country has made a significant breakthrough by replacing imported copper through indigenous aluminium.

The country spends about 300 crores per year on the import of spare parts. This difficulty has been accentuated because of the large-scale imports of machineries and thinking only in terms of part number and not the function of component through codification process. This process has resulted in industries carrying Rs 2,500 crores of obsolete (mostly imported) spares. While importing the capital machinery, adequate precaution should be taken to buy the drawings as well.

It is too well known that the lead time for the clearance of import requirements is very long. Further, the requirements of imported material are to be purchased in annual or six-monthly consumption quantities. These factors compel all organisations to carry a high inventory of imported materials. In spite of several procedural changes introduced recently by the DGTD, the senior materials executives attending Administrative Staff College courses are of the firm view that there is plenty of scope for reduction of lead time in DGTD and CCIE.

These Central organisations should also take initiative in setting up "Banks" for critical imported raw materials and the spare parts. Obviously this will reduce not only the imports at the macro level but also investment on the inventory at the micro level.

Role of R&D in Import Substitution

Technological innovation is essential to the birth and growth of industrial enterprises and the economic, social and political health of nations. The stock of technology on which innovation depends can be increased by Research and Development. It is neither necessary nor practical to rely entirely on one's own Research and Development and it would appear that every country must buy technology from abroad. But to make the best of the bought-out technology it needs an internal Research and Development effort to adapt and to absorb. Therefore, the strategy for achieving technological self-reliance must include: (1) Selective buying of technology, (2) Import substitution, (3) Development of threshold technology, and (4) Development of essentially indigenous technology.

CSIR. The agencies in our country that are supposed to look after the development of indigenous technology as well as keep pace with the technical developments in advanced countries, are the laboratories of CSIR—Council of Scientific and Industrial Research—and the unit level R&D. The CSIR has a chain of forty laboratories covering almost all the fields of technology. Some of the laboratories have been in existence for more than a quarter of a century. The annual budget is about Rs 30 crores. And over 10,000 technical and scientific personnel are employed in them. Admittedly, a few of these laboratories have done commendable work; but, unfortunately, the measure of the output from CSIR laboratories, according to information from NRDC—National Research Development Corporation—adds an additional 10 crores only to the gross national product (GNP). This is also corroborated by the data on the import substitution awards given by the inventions promotion board which is given in Exhibit 16.4.

EXHIBIT 16.4

ANALYSIS OF IMPORT SUBSTITUTION AWARDS FOR 1970–75

Agency	*Per cent of awards won*
Research laboratories*	29
Public sector industries	16
Private sector industries	48
Freelance inventors	7

**Note*: The research laboratories include the defence, agricultural, medical, atomic energy and CSIR laboratories. Consequently the share of CSIR will be much less than indicated above.

Our experience in conducting courses for scientists from the laboratories as well as from industries in Administrative Staff College of India indicates that better return on investment on research will accrue if the laboratories are associated with industry groups and made responsible to them. In this context the objectives of the laboratories need a close look and should be brought in line with the industrial development plan of the nation.

Similarly the tax-payer would like to know the contribution of 10,000 technical personnel in 37 defence research laboratories with an annual budget of Rs 35 crores, towards import substitution.

Unit level R&D. Individual organisations, instead of waiting for the Government to introduce the above suggested change, should take the initiative to establish their own R&D. Obviously at the unit level, design and development will be more productive. In this context, the concept of research surcharge, which has been floating around for the last four years, is yet to be implemented. However, the industry can emulate the example of BHEL, which has set up a separate corporate R&D unit. Though the National Committee on Science and Technology (NCST) has recommended that one per cent of the Gross National Product (GNP) should be spent on R&D efforts, progressive companies like BHEL, BEL, etc., are spending more than three per cent of their sales on R&D efforts. Small units can invest in R&D by also sponsoring projects in universities and IITs in order to properly utilise the talent of over one lakh scientific and technical personnel in the country. (Reference page 21 of Research and Development Statistics 1973–74 published by NCST.)

Technology transfer. The difficulties in the transfer of technology from research institutions to shop floor in the country, are too well known to be enumerated. This obviously arises due to lack of communication and the inability to understand the constraints of a production organisation by the scientists and vice-versa. In this context our earlier suggestion of having laboratories that are responsible to industry groups, assumes great significance. Further, we piously hope that the idea of the polytechnology clinic, which is being promoted by CSIR, will be a step in the right direction.

As mentioned earlier, all countries to some extent have to import technology. This has a dynamic role to play in a developing country like ours, but only when dovetailed to and closed-knit with other technologies in our reckoning. Having amassed imported technology over the past two decades, the quickest pay-back to our economy is from adaptive technology or import substitution. Development of essentially indigenous and threshold technology, however, has to be the bulk of our effort not only to achieve self-reliance, but also to build up export potential. However, for better assimilation of imported technology, it should be purchased outright or a dynamic research institution has to be associated with every import of know-how.

Technology transfer centres. In order to achieve the above, the government should set up "technology transfer centres" for conducting techno-economic reviews of the old and new patents field in India. Such technology transfer centres can pick up some of the promising inventions for commercial exploitation after input of requisite R&D effort. While lapsed patents may be utilised freely, in the case of current patents, provision of compulsory licensing under the patent laws may be availed of to break the foreign monopoly hold on the existing know-how.

The same centre may also conduct techno-economic review of foreign patents and allied publications of industrially advanced countries, classify promising ideas industry-wise, and feed the same to appropriate R&D institutions and industries. This will provide them with the necessary starting point for marginal R&D work needed for an import substitution effort, development of threshold technology, improvements in production methods, as well as diversification of production to stimulate growth of new industries.

Problems in Import Substitution

In the process of import substitution, companies and various central agencies, particularly the **DGTD**, encounter a series of bottlenecks. Some of these are illustrated below:

1. Authorities in charge of execution of projects—whether in public or private sectors—tend to give very tight specifications and standards. These critical specifications make the task of mobilising the manufacture of such equipments from indigenous resources very difficult. Normally many of the specifications are borrowed and the relevance of such tight specifications for Indian conditions should be reviewed.

2. Project authorities ask for imported capital equipment on the strength of absence of quality assurances or performance guarantees from the indigenous machine building sector.

On the one hand, it is difficult in the context of early commissioning of projects to resist such claims, on the other, it is equally necessary that sufficient time, based on advance and coordinated planning, is given to indigenous machinery manufacturers to be able to satisfy the quality and performance aspect. It is DGTD's experience that unless there is a start with development contracts, indigenous manufacturers rarely come up.

3. There is a large communication gap *inter se* between industries in a situation where the output of one industry is the input of another. This gap results in pressures for import.

4. Engineers and users are accustomed to part numbers, usually of imported components, and are reluctant to go into development of drawings and to question the specifications.

5. Sometimes import substitution becomes difficult in view of the high image of foreign brand names amongst the elite consumer in India—Coca Cola is a case in point.

6. The foreign collaborators would like to keep the Indian licencees and subsidiaries indefinitely dependent upon their raw material, components and know-how.

CONCLUSION

There is a growing awareness at all levels for self-sufficiency and import substitution. The responsibility of this task is shared by the government and the industry. Industry should take a constructive attitude towards import substitution and shed its inhibition for a change. At the micro level, in the industry the responsibility of initiating import substitution rests on the materials manager who is incharge of source development. However, he has to get the cooperation of his colleagues for technical aspects, quality, quantity, price, etc. Scientific laboratories like CSIR should take an active part in the process and aid the industry with sufficient information and testing facilities.

The facilities and expertise already established in the country have to be availed of to full advantage in making more exports, fuller utilisation of capacities, materials conservation and import substitution, coupled with a curtailment of non-essential imports catering particularly to elite consumption. It is hoped that the country with the available infrastructure will achieve greater import substitution and reduce the trade deficit in the near future.

17

Public Buying

THE DGS&D

The centralised buying concept has rightly gained credence because it helps to keep wasteful competition and duplicative efforts to a minimum. The Directorate General of Supplies and Disposals constitute the central purchasing organisation of the Government of India. It is responsible for the procurement and purchase of supplies required by the indenting departments of the Government of India, including the defence organisation, railways as well as posts and telegraphs departments. Besides, its services are also utilised by the State Governments, Public Sector Enterprises and Quasi-public bodies. The successive five-year plans have considerably increased the volume and variety of purchases made by DGS&D. The centralised purchasing organisation has enabled a considerable degree of standardisation of the procedures for tendering and contracting. The DGS&D, being a government department, is accountable to the Parliament for its action and this perhaps is the reason for its tender enquiries and contracts cluttered with far too many points.

Further, DGS&D scrutinises proposals for the purchase of stores from abroad with a view to ensure encouragement of Indian industries and to prevent, as far as possible, purchase of stores from outside India when goods of suitable quality are indigenously obtainable. It inspects stores purchased direct by Government Departments and other bodies when the purchasing departments want to make use of the services of DGS&D Inspection Wing. It carries out laboratory tests, analysis, etc., on behalf of those authorities who want to make use of the services of DGS&D in this respect. It arranges for shipment of stores exported and clearance of stores imported. Further, it is also responsible for the disposal of surpluses declared by Central Government Departments and State Governments.

Volume and Nature of Purchases

Over the past decade, DGS&D has steadily increased the value and variety of its purchases. Although the majority of the purchases are made within the country, DGS&D also imports. It has its subsidiaries in London and Washington to coordinate all its imports. The imports from the East European countries are based on the trade agreements with them. The DGS&D charges ½ per cent on the value of purchases as service fees and another ½ per cent for inspection purposes.

An idea of the growth of the activities of DGS&D can be had from the fact that the annual average purchase value rose from Rs 450 crores during the Third Plan period—i.e. 1966—to about Rs. 1,000 crores* during 1975. As much as 90 per cent of these purchases were made from Indian sources which have been gradually increasing every year. It is, therefore, essential that a sound organisational set-up be provided to ensure effectiveness.

Organisation of DGS&D

The detailed organisational arrangements are shown in Exhibit 17.1. The organisation is headed by the Director General, assisted by one Additional Director General and five Deputy Directors General. The regional offices are located in Madras, Bombay, Kanpur and Calcutta. DGS&D has separate wings for supplies, inspection, progress and disposals. The working of the DGS&D has been reviewed in detail by the estimates committee of the Parliament, which has made valuable comments.**

The Supply wing is the mainstay of the organisation and has 11 directorates at headquarters and 5 regional purchase directorates in other parts of the country. The grouping of the purchase directorates has been done so as to ensure that homogeneous groups of items are dealt with at one place as far as possible. The directorates at the headquarters only handle indents of centralised items. The planning and development division helps to make the procurement process a success by rationalising specifications, and drawing a comprehensive programme for receipts and booking of indents of common user items. The O&M Branch assists the supply wing in devising procedures, systems and effective control measures to maintain optimum efficiency. The Statistical branch which has been computerised recently is responsible for maintaining up-to-date statistics of the requirements and publishes the *Annual Report* as well as *Directory of Government Purchases.*

The inspection wing under the charge of the Deputy Director General (Inspection) ensures that the stores obtained from the suppliers are of the right specifications for the needs of the indentors, and also monitors quality during the manufacturing process. The wing also advises the indentors on technical matters and on the suitability of the particular stores for their requirement. For this purpose, this organisational set up is spread all over the country.

The progress wing under the charge of a Deputy Director General is responsible for progressing all contracts placed by DGS&D. It keeps a close watch on the supply position after the placement of contract and locates the cause for delay in supply, through the field officers. This wing has been divided into three branches, viz., Defence, Railways and General. With the help of field officers, vigorous follow up is done on behalf of the respective indentors. It also routes supplies where urgently needed. To ensure expeditious supplies of critical items to major indentors, periodical meetings are held to consider issues like coverage of indents, supply position of contracts, and bottlenecks which may hold up supplies.

* 1 crore = 10 million.

** Estimate Committee, 121st Report, Lok Sabha Secretariat, Government of India.

EXHIBIT 17.1

ORGANISATION CHART—DIRECTORATE GENERAL OF SUPPLIES & DISPOSALS

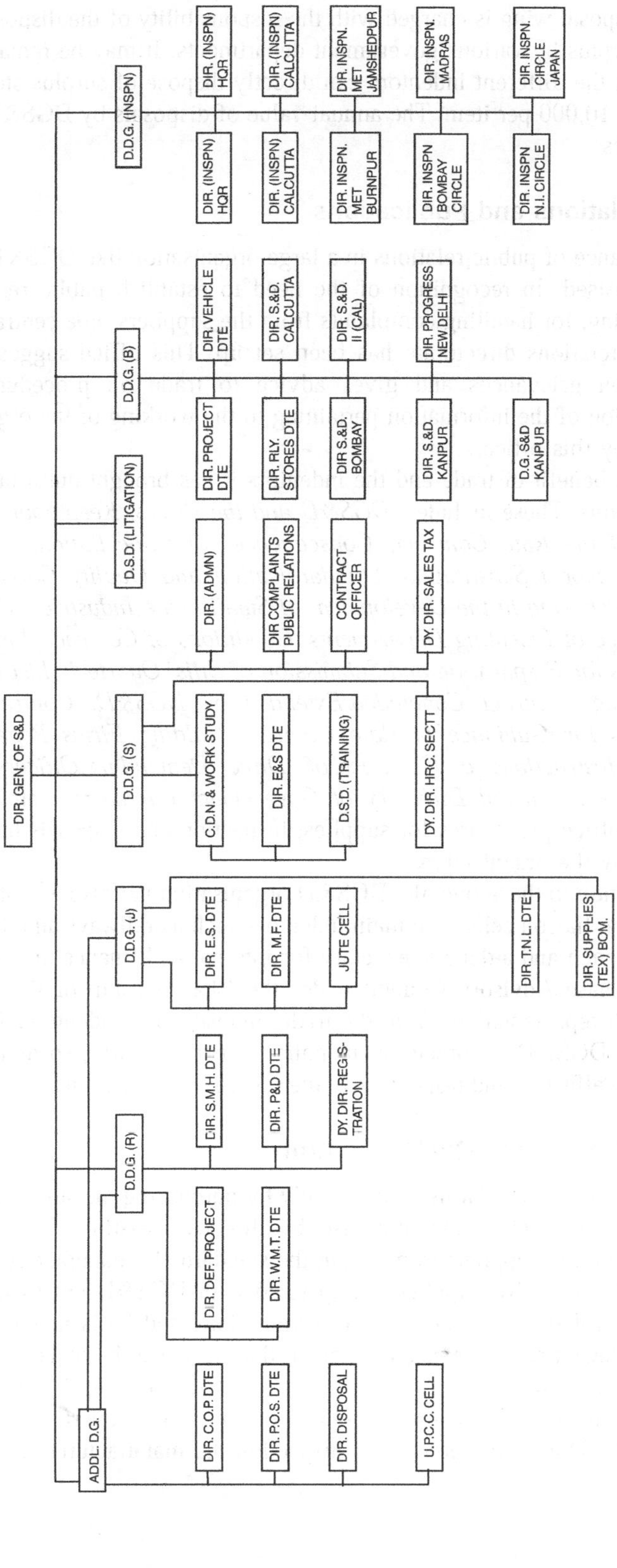

The disposal wing is charged with the responsibility of the disposal of stores declared surplus by various government departments. It may be remarked in this context that the different indentors can directly dispose of surplus stores up to a value of Rs 10,000 per item. The annual value of disposals by DGS&D is around Rs 50 crores.

Public Relations and Publications

The importance of public relations in a large organisation like DGS&D cannot be over-emphasised. In recognition of the need to establish public relations on a proper footing, for handling complaints from the suppliers on a centralised basis, the public relations directorate has been set up. This office suggests remedial measures on grievances and gives advice to trade on procedural matters. Dissemination of the information pertaining to the working of the organisation is also done by this office.

For the benefit of trade and the indentors it has brought out a large number of publications. These include *"DGS&D and the Prices, Keep Your Date, Make the Best of the Rate Contract, Consequences of Cancellation, Let Us Help Ourselves, Import Substitution, Standardisation and Quality Control, Role of Central Purchasing in the Development of Small Scale Industries, Pamphlet for the Guidance of Indenting Departments, Conditions of Contract, Instructions to Contractors for Preparation and Submission of Bills, Quarterly List of Stores on Rate/Running Contract Concluded/Extended by DGS&D, Contract Manual, Instructions for Guidance of Rate Contract Holding Firms Regarding Rate Contracts, Instructions for Guidance of Direct Demanding Officers Regarding Rate Contracts, Annual Directory of Government Purchases"*, etc. A detailed manual of office procedures for supplies, inspection and disposals has also been published by the organisation.

In addition to the above, the DGS&D organisation has also a contract officer, legal adviser, liaison cells with main indentors such as railways and defence, cost accounts branch and a deputy secretary for internal and finance matters. There is also a Purchase Advisory Council under the Chairmanship of the Minister of Supply with representatives from the trade, industry and indentors. This council advises the DGS&D on procurement policy and procedures, so as to make the organisation effective and responsive to the socio-economic changes in the country.

Standardisation and Quality Control

DGS&D promotes standardisation not only by purchasing the stores according to some standard specification, but also by directly assisting, encouraging and insisting upon the suppliers to maintain the standards in manufacture and quality control techniques. No supplier is registered with DGS&D unless it is verified that he has got the necessary technical know-how and he is manufacturing the stores to relevant standard specification so that these can be objectively tested at an independent laboratory. Again, after the orders are placed, all supplies are tested consignmentwise and only those conforming to the governing specifications are accepted. During the process of inspection the manufacturers are also given technical advice, wherever necessary, for improvement in quality control. Complaints against the supplies arranged through the DGS&D are as low as two per thousand or 0.2 per cent.

Development of Indigenous Sources of Supplies

DGS&D in collaboration with the Directorate General of Technical Development makes efforts to locate indigenous sources of manufacture. In respect of imported spare parts and components for the maintenance of earth-moving and construction equipment, about 15,000 items of indigenous spares and components have been located and brought on rate contracts. In respect of non-sophisticated items the users are advised to accept alternative items which are available indigenously and which can serve the functional end. The percentage of imported stores in relation to total purchase has come down to 8 per cent during 1974–75 from 24 per cent during 1967–68.

Purchases from Cottage and Small Scale Industries

The Central Purchase Organisation is required by Government Policy to stimulate the growth of the small scale industrial sector. 167 items of stores are reserved for exclusive purchase from the small scale sector, and more items are considered from time to time for this kind of preference. A price preference to small scale industrial units is allowed vis-a-vis offers from non-small scale industries. The volume of purchases made from the cottage and small industries has been steadily increasing.

Registration of Suppliers

All firms dealing in stores purchased by DGS&D, whether they are in India or abroad, are eligible for registration. The firms are registered in the capacities of manufacturers, stockists of imported stores and certain indigenous stores, e.g. textiles, timber, etc., sole selling agents of Indian manufacturers and agents of overseas principals. More than 4,000 suppliers are registered with DGS&D. A supplier is asked to file an application. The bona fides of the supplier including the equipment, plant and machinery, technical capacity and funds position are verified partly through enquiries on the spot by the officers. The vendors are also rated on the basis of delivery, quality, price and reliability.

Firms registered with DGS&D enjoy the following advantages:

1. Demands which are not advertised are normally met from registered firms.
2. In the case of advertised demands, copies of the notices are sent to registered firms, giving them advance information of the demand.
3. Rate contracts and running contracts are normally reserved for registered firms.
4. Operational priority demands are normally reserved for supply by registered firms.
5. Normally 80 per cent of the quantity against express and urgent demands is reserved for supply by registered firms. Even ordinary demands are mostly covered from registered suppliers.

6. Registered firms are not required to furnish security deposit.
7. The State Governments and Public Undertakings may at their discretion waive the requirement of security deposit from firms registered with DGS&D.
8. Registration with DGS&D raises the firm's standing in business not only with Government departments but also with foreign firms and establishments.

Rate and Running Contracts

The Central Purchase Organisation purchases all stores and standard types which are in common and regular demand and the prices of which are not subject to appreciable market fluctuations, on the basis of Rate or Running Contracts, whichever is most suited to the circumstances of each particular case. Normally, a Rate/Running Contract is concluded for a period of one year, and only with firms whose capacity to deliver the goods to the requirement of the indentors has already been verified. The user is thus assured of a steady flow of supply as and when he needs the stores. He is, therefore, freed from the necessity of always maintaining large stocks of numerous items which would lock up his funds as well as block his available storage space. The tedious and time-consuming process of locating the cheapest available source of supply of goods of acceptable quality every time he needs a particular item, is eliminated. The suppliers, on the other hand, are assured of a reasonably large demand, spread over a fairly long period, derive benefits from comparatively more economical planning of their production programmes, and continuous technical advice from the expert inspection organisation of DGS&D. Over a period of years, DGS&D has decreased its dependence on "open tenders" and an increased number of contracts—about 80 per cent—are finalised on the basis of limited tenders and negotiations. More often than not tender enquiries are addressed to people whom the DGS&D know as benign suppliers or who have in the past supplied satisfactorily. Such limited enquiries are preferred to avoid the time lag necessary for advertisements.

Payment Procedure

Against DGS&D contracts, the Trade & Industry receive advance payments to the extent of 90/95/98 per cent as the case may be, immediately after despatching the inspected stores to the consignee. The suppliers' bills are paid by the Pay & Accounts Officer within 14 to 21 days of their presentation if they are otherwise in order. By adopting this system of advance payments it is ensured that the capital of the Trade is not blocked and that they receive the major portion of their payment immediately on completion of supplies.

Indent Processing

On receipt, the indent is registered in the Central indent section. If the specification requires checking, it is sent to the inspection wing for thorough scrutiny. Later, it is passed to the Tender Enquiry Section of the Purchase Directorate concerned, for initiating procurement action. Wherever a difficulty arises, the indent is referred back to the indentor. The invitations to tender are issued through advertisement, limited tender enquiries or single tender basis, depending on value, nature and urgency of the requirement. Normally DGS&D takes on an average about three months between the date they have an indent on which action can be taken and the date on which a contract is placed. Once a contract is placed, the supplier informs the inspectors as soon as the item is ready for inspection. The inspected stores are then despatched to the consignee by the supplier. The supplier normally takes a month or two to commence supplies. Then comes the transit time. The effort is always to see that the actual supply period and its completion meet the requirements laid down by the user in his indent. However, only 50 per cent of the contracts are completed within the original stipulated period. About 70 per cent of the contracts are completed with extensions up to six months. In this context it may be mentioned that DGS&D ensures availability of critical raw materials to the suppliers. For instance, DGS&D, on an average, provides about 150,000 tonnes of steel of one kind or the other to various suppliers. The lead time taken by the DGS&D has always been criticised by the indentors and suppliers. In this context the estimates committee of the Government of India in its *121st Report* has recommended that a relentless effort should be made to simplify and streamline the procedures so that action can be taken expeditiously at every stage for allaying the apprehensions of the suppliers and the demand of the indentors.

CONCLUSION

The DGS&D, as the Central Purchasing Organisation of the Government of India, has proved a very valuable organisation. To quote Estimates Committee in this context: "The Committee also observes that the organisation has stood the test of time and has shouldered the responsibility cast on it in regard to the maintenance of supply lines in the country having due regard to the development of Indian industries both in times of peace and war and according to the exigencies of the situation resulting from emergencies from time to time."*

*Estimates Committee: 121st Report, Lok Sabha Secretariat, Government of India, page 7.

18

Legal Aspects in Buying

The materials manager commits a good deal of corporate finance when he places purchase orders. It is imperative that he understands the legal aspects that are relevant to his profession. The Indian Sale of Goods Act of 1930 and the Indian Contract Act of 1872 cover some of the important legal aspects.

While a materials manager may not be well conversant with all branches of law, at least his profession demands that he possess a working knowledge of some of the important laws, which are essential, namely (i) Law of Agency, (ii) Law of Contract, (iii) Law pertaining to sale of goods, and (iv) Arbitration.

LAW OF AGENCY

We live in the area of trans-nations in which business intercourse is growing heavier and heavier. As such, the multiplicity of business transactions enjoins upon a businessman to depend upon the services of others. In legal parlance, normally such persons who provide services are called *Agents.* The law pertaining to the relationship between the Agent and the Principal is defined in the Law of Agency.

Definition of Agent

An Agent is a person employed to act on behalf of another called the Principal, for the purpose of bringing the Principal into contractual relationship with a third person. A person who has the capacity to contract may enter into a contract with another either by himself or through another person. When he adopts the latter course, he is said to be acting through an agent, and the person for whom such an act is done or who is so represented is called the *'Principal'*. A Contract of Agency may be made verbally or in writing. When it is in writing, it is in the form of Power of Attorney.

Rules of Agency

Whatever a person competent to contract, may do so, by himself or through an agent. "Qui facit per actum facit per se". This means, 'he who does through another, does by himself or the acts of an Agent subject to certain conditions, are the acts of the Principal'.

Essentials of Relationship of Agency

The two essentials of relationship of Agency are Agreement between the Principal and Agent and Intention of Agent to act on behalf of the Principal.

Agreement. Agency depends on agreement, but not necessarily on contract. No consideration is necessary to create an agency. The fact that the Principal has agreed to be represented by the Agent, is sufficient 'determinant' to the Principal to support the Contract of Agency.

The relationship of Principal and Agent may be created by the express agreement and by implied agreement.

Agency by express agreement. A contract of Agency may be created by an express agreement. Normally, the authority given by the Principal to his agent binds the Principal by the acts done within the scope of his authority. The usual form of a written contract of Agency is the Power of Attorney on a stamped paper.

Agency by implied agreement. Agency is not necessarily contractual in nature. The existence of Agency may in certain cases be implied when it is to be inferred from the circumstances of the case. For example, 'A' owns a shop in Hyderabad, himself living in Secunderabad. The shop is being looked after by 'B' and during the course of business, he buys ordinary goods from 'C' for the purpose of running the shop, and pays for them out of 'A's funds with his knowledge. Here, 'B' has implied authority from 'A' to order goods from 'C' and make payment in the name of 'A'.

The Implied Agency arises when the Principal conducts himself, towards the agent or the third parties, in such a manner as if the Principal had conceded to the appointment of the Agent. Implied Agency includes Agency by Estoppel, Agency by Holding out, and Agency by Necessity. A detailed study of such agencies is beyond the scope of this text.

LAW OF CONTRACT

From the Materials Manager's point of view, the Law of Contract is an important branch of Mercantile or Commercial Law, because all commercial transactions start from an agreement between two or more people. The object of the Law of Contract is to introduce definiteness in commercial and other transactions. It may be said that the purpose of the Law of Contract is to ensure the realisation of reasonable expectation of the parties who enter into a contract. The Law of Contract may be defined as that branch of law which determines the circumstances in which a promise shall be legally binding on persons making it. The Indian Contract Act of 1872 (Act IX of 1872) lays down certain general rules regarding contracts. The act is not exhaustive. There are other acts relating to particular types of contracts, e.g. The Negotiable Instrument Act, The Transfer of Property Act, etc. However, from the purchasing agent's point of view, the Law of Contract is more important.

Section (6) of the Indian Law of Contract provides that 'An Agreement enforceable by Law' is a contract. The Law of Contract differs from other branches of Law in an important respect. It does not lay down a number of rights and duties which the Law will enforce; it consists rather of a number of limiting principles subject to which the parties may create rights and duties for themselves which the Law will uphold. The parties to a contract in a sense make the law for themselves. So long as they do not infringe some legal prohibition, they can make what rules they like in respect of subject matter of their agreements.

Definition of Contract

A proposal or an offer when accepted is a promise. A promise and every set of promises forming the condition for each other is an agreement. An agreement, if made with free consent of parties competent to contract, for a lawful consideration and with a lawful object, is a contract. A contract has been defined as 'an agreement enforceable by law', or, in other words, 'a legally binding agreement between two or more persons by which rights are acquired by one or more to act or forbearance on the part of the other or others'.

If we analyse these definitions, we find that a contract consists of two elements: an *agreement* and its *enforceability* by law. An agreement is the same as a promise. A promise is defined thus, "when the person to whom the proposal is made signifies his assent thereto, the proposal is said to be accepted. A proposal when accepted becomes a promise. Hence, in order to form an agreement, there must be an offer or proposal by one party, and the acceptance of that proposal by the other. To sum up: Agreement = Offer + Acceptance.

An agreement connotes that at least two parties expressed themselves with sufficient certainty, and in terms which correspond, on the subject.

Agreement is a very wide term. If 'A' invites 'B' to a dinner, and if 'B' accepts the invitation, it is a social agreement. Social agreements are not contracts as they are not enforceable under the Law of Contract. It is only agreements enforceable in Law courts, which are contracts.

To conclude, Contract = Agreement + Enforceability at law. All contracts therefore are agreements, but all agreements are not contracts.

An agreement may be made in writing, by word of mouth, by reference from the conduct and circumstances of the case; or by any combination of the above methods.

Essential elements of a valid contract. An agreement is a wider term than a contract, because an agreement may or may not be enforceable by law. It is extremely important to know what agreements are contracts in Law. In respect of an agreement which is not a contract, no legal remedy is available to either party, if the other party fails to carry out the agreement.

A contract, in order to be enforceable by Law, must have the following essential elements:

1. *Offer and acceptance.* There must be a lawful proposal or offer by one party and a lawful acceptance of that proposal by the other party thus resulting in an agreement. The terms of offer and acceptance must be definite. The

acceptance of the offer must be according to the mode prescribed and must be communicated to the proposer. A proposal when accepted becomes a promise.

2. *Legal relationship.* The agreement must create legal relationship between the parties. If there is no intention to create legal relationship, there can be no contract between the parties.

3. *Lawful consideration.* The agreement to be enforceable by law, must be supported by a consideration. The term consideration in simple words means 'something in return'. The agreement is legally enforceable only when both the parties give something and get something in return. A promise to do something, getting nothing in return, is usually not enforceable by Law.

4. *Capacity of parties—Competency.* The parties to the agreement must be capable of entering into a valid contract. Every person is competent to enter into a contract if he is of sound mind, age of majority, and not disqualified from contracting by any law to which he is a subject. Flaws in capacity to contract may arise from minority, lunacy, idiocy, drunkenness and status.

5. *Free and genuine consent.* It is essential to the creation of every contract that there must be free and genuine consent of the parties to the agreement. The parties must agree upon the same thing in the same sense, i.e. they must be of the same mind on all the material terms of the contract.

6. *Lawful object.* The object of the agreement must be lawful. It is lawful unless it is forbidden by law or is of such a nature that if permitted, would defeat any provisions of law.

7. *Agreement not declared void.* The agreement, though it might possess all the essential elements discussed above, must not have been expressly declared void by any law in force in the country.

8. *Certainty and possibility of performance.* The terms of the agreement must be certain and not vague or ambiguous. If the agreement is vague and if it is not possible to ascertain the meaning of agreement, it cannot be enforced. The terms of agreement must enable performance. An agreement to do an act impossible in itself cannot be enforced.

9. *Legal formalities.* The agreement may be oral or in writing. Where it is in writing it must comply with the necessary legal formalities as to writing, registration and attestation. If the agreement does not comply with these legal formalities it cannot be enforced by law.

Contract of Sale

When an offer to sell or buy goods for a price is made, it becomes a contract of sale. However, it is binding only when accepted by the buyer and his acceptance is intimated to the seller on the terms and conditions on which it is made. Also any change in terms and conditions is equivalent to a rejection of the original offer. In other words, it becomes a counter offer. On the other hand, an enquiry requesting whether the tenderer could modify his terms does not make the contract void.

Aspects in Proposals and Acceptance

It is common practice in the industry to place orders through 'telephones'. In order to make this binding, a formal purchase order should follow. A proposal cannot be revoked once the acceptance is put in a course of transmission to the proposer. In a similar manner the acceptance cannot be revoked once the communication of the acceptance comes to the knowledge of the proposer. Suppose, Ram proposes to sell an item to Rahim by a letter sent by post. Rahim accepts the proposal by a letter of post. Ram can revoke this proposal at any time before Rahim posts his letter of acceptance but not afterwards. Rahim can revoke his acceptance at any time before his letter reaches Ram but not afterwards. However, the party who gives time to another to accept or reject an offer is not bound to wait till the time expires.

Special Contracts

Quite often materials managers place contracts where a rate is agreed upon and the quantities to be supplied from time to time are left to be intimated by periodical releases. Here, the buyer does not commit himself to lift any minimum quantity. Such a contract is deemed to be only a standing offer and its acceptance only means that the buyer recognises this standing offer for the stated period. However, once a release order is issued by the buyer the contract becomes binding for that quantity. The supplier may feel free to withdraw his standing offer with reference to future orders. This is not possible if the tender and its acceptance were intended by the buyer and the seller to be a contract involving an obligation on both of them during that time period.

Frequently contracts are entered into, stipulating a specific quantity to be lifted during a stated period, with clauses providing some variations in the quantity or the period during which it is to be lifted. This is a binding contract and enforceable for the supply of the entire quantity of goods contemplated in the contract. There is another type of contract in which, though the buyer is not bound to any specific quantity, he commits to buy and pay for all the goods needed by him or goods that are produced by the seller in a stipulated period. This again is a binding contract. A purchase contract becomes null and void when the price to be paid is not stipulated in the order or at least the manner in which the price is to be fixed is not indicated.

However, according to the Indian Sales of Goods Act, stipulations regarding time of payment are not conditions of the contract. In other words, they are not of essence. This is because a delay in payment does not make the contract without consideration. But if the contract is for the delivery of goods by instalments and each instalment is to be paid for on delivery, then failure to pay for any one instalment is adequate for the seller to repudiate the contract, when the time of payment for each instalment is of the essence. In determining whether the time was of the essence of the contract the court will consider the intention of the buyer and the seller as expressed in the contract as well as in the course of dealing between the parties or usage, surrounding circumstances, equitable considerations, etc. Ordinarily, however, when a specific time is fixed, it is treated to be of essence and the onus of proving that it is not so lies with the one who asserts so. To quote Indian Sales of Goods Act, "Unless a different intention appears from the terms of the contract, stipulations as to the time of payment are not deemed to be of the essence of a contract of sale. Whether any other stipulation as to time is of the essence of the contract or not depends on the terms of the contract".

Condition and warranty. A condition is a stipulation in contract of sale pertaining to goods which are the subject of contract and this stipulation is essential to the main purpose of the contract, the breach of which will make the contract liable to be rejected. A warranty is also a stipulation in contract of sale pertaining to goods which are the subject of contract *but* this stipulation is only of a collateral nature to the main purpose of the contract, the breach of which gives rise to a claim for damages but does not make the contract liable to be repudiated and hence the goods cannot be rejected.

Sale by description. It is not uncommon to find orders which have only a description of the item in question. By implication, it follows that items must follow the description. It pays in the long run to make certain that such purchase orders are complete and clear in their description. This could be done by mentioning the end use, furnishing drawings, tolerances, finish, etc., referring to the national or international specifications, and the tests which the item has to pass.

There are cases when the supplier produces the item under the full knowledge as to how exactly it will be used. In such a case there is an implied warranty in the sense that the buyer can claim damages when the item later proves to be unfit. There is a mistaken notion that there is an implied condition in the case of a contract for the sale of a specific item under its patent. There is actually no implied condition as to the fitness of the item for any specific use. Similarly, statements regarding the quality of the product and performance appearing in commercial advertisements are not to be construed as an express or implied warranty.

Aspects in contracts based on samples. Goods sold on the basis of samples give the buyer many advantages. The buyer must be able to establish the existence of a sample as well as a clear understanding to the effect that the sample is of the contract standard. For this purpose the sample needs to be preserved for future inspection.

Works contract. This type of contract pertains to the supply and fixing of materials to buildings and such other immovable properties. The important factor in works contract is that it is indivisible, i.e. it cannot be looked upon as a contract of sale of goods plus a contract for labour. From the materials manager's point of view a good amount of saving can result by way of reduction in sales tax through resorting to works contract.

We will now see the difference between a guarantee and warranty. A warranty is a part of the contract though only collateral to the main object of the contract. It is a statement of something which a party undertakes to do. A guarantee is something which the promiser undertakes, to be answerable to the promisee for the debt, default, etc., of another person whose primary liability to the promisee himself exists or is contemplated. Therefore, it is an additional and conditional contract.

We will discuss the legal aspects involving the payment of price. It has three dimensions, viz., time of payment, place of payment and mode of payment. As regards the time of payment, the amount should be paid when it is due. The buyer has no right to delay it till a demand is made. As regards the place, it should be the place previously appointed. If unspecified it should be at the seller's place. The mode of payment is very important. In the purchase order, this should be clearly spelt out. Factors like sales tax, octroi, customs duty, excise duty, escalation classes, discounts, credit terms, despatch terms, such as FOB, CIF, etc., must be clearly laid down. If the buyer complies with the mode of payment suggested in the contract, he is safe even if the money is stolen or does not reach the seller.

Risks and rights. It is necessary for the buyer to recognise that the risk of deterioration of goods during transit rests with him (unless of course the deterioration is abnormal) even though the seller might have agreed to deliver the goods at his own risk at a place other than where they are when sold. The buyer, of course, has his rights. Unless he has a reasonable opportunity of examining the goods delivered, it will be deemed that they have not been accepted. However, if the buyer fails to inspect the goods within a reasonable time and determine whether they conform with contractual terms, then he is deemed to have waived his right. If by mistake the buyer has failed to discover defects then subsequently he has a right to claim damages.

Acceptance of goods. Acceptance by the buyer is complete only when he has intimated the seller of his acceptance, or when he fails to intimate the seller even after a reasonable time, during which he had opportunities to inspect the goods, or he does not act in relation to them which is inconsistent with the ownership of the seller. The buyer has another important right. Unless agreed upon, the buyer is under no obligation whatsoever to return the goods which he has rejected as they do not conform to contractual requirements.

Breach of Contract

The party which sustains a loss due to breach of contract will be so compensated with respect to damages that it will be placed in a position as if the contract has been performed. In commercial transactions the aggrieved party can claim only the actual loss caused to him and not to others.

Damages arising out of a breach of contract in the usual course are recoverable. However, damages which arise owing to circumstances peculiar to the case and which would not have arisen during the normal course are *not* recoverable. Unless the circumstances are made known to the party which breaks the contract and such damages are contemplated by both the parties at the time of entering into the contract. The amount of compensation by way of repatriation of damages will be, as far as possible, the amount which will place the aggrieved party in the same position as it would have been, had it not suffered the breach of contract which led to such damages. Very often the parties wish to reduce damages to a certainty. This, they achieve by fixing a rate or sum at which damages are to be assessed. If such an amount is a genuine pre-estimate of the damages, then it is known as liquidated damages. However, if the amount thus fixed is unconscionable with respect to the greatest loss that could have followed from the breach, it would be considered a penalty. As per the Indian Contract Act, it is for the court to ascertain the actual loss and award the same. A similar provision is available for security deposit and earnest money. If the amount is small and in reasonable proportion to the price then the forfeiture of the deposit or earnest money paid by the defaulting party is not considered as penalty. If the amount is a large part of the price, then forfeiture of the amount will be in the nature of penalty. In such cases the Court will not permit forfeiture but will only award reasonable compensation.

Patent rights. The buyer can protect his organisation by adding special patent infringement clauses to his purchase orders. However, when the buyer orders to his own specification, this does not apply. It is often the case that big buyers give financial and technical assistance to the supplier for developing new products and materials. It is advantageous, in such cases, for the buyer to include a clause in the purchase orders, covering the patent rights in inventions discovered by a supplier in the course of executing his purchase contract.

Settlement of disputes. Arbitration is used for settling disputes between buyers and sellers. It is less time-consuming when compared with litigation. For this purpose the buyer must include an arbitration clause and the supplier has to agree to it. Once such a clause is stipulated, the court will not entertain any suit unless the arbitration process is exhausted. Arbitrators have similar powers as the courts and their award is binding and enforceable on the parties.

CONCLUSION

Professional legal advice is a must for major contracts involving large sums of money or extending over a long period of time like in imports. Contracts should be standardised after scrutiny by the company's legal department. The International Chamber of Commerce is the arbitrator for resolving disputes arising between parties belonging to two different countries. The ICC has handled more than 300 arbitrations. ICC has also codified general business terminology to minimise friction through different interpretations.

In the context of legal buying, it may be worthwhile to recapitulate the slogan of the Liverpool Stock Exchange which says, "my word is my bond". In spite of all legal precautions taken by the materials manager, there are bound to be some loopholes. Hence, after taking sufficient precautions in the contracts, emphasis should be on mutual trust and confidence.

19

Insurance Buying

RESPONSIBILITIES OF THE MATERIALS MANAGER

The materials manager's association with insurance begins from the time he places a purchase order for incoming materials, continues during the period the goods are stored at the plant, and concludes only when the finished goods reach the hands of the ultimate user.

The materials manager has to arrange insurance of materials in transit and in stores scientifically, so that the possibility of catastrophic and profit shattering losses either to the organisation or to those dealing with the organisation, like the suppliers and purchasers, is avoided. As controller of movement of materials he has to ensure that claims, if any, against the insurance are lodged at the proper time and in such a way that the compensation from insurers is received in full and with minimum delay. Further he has to minimise wherever possible the cost of insurance consistent with the safety of the goods. The cost of the insurance in industry varies from about 1.5 per cent of the ultimate cost of the product to about 10 per cent in some cases, with an average cost of about 3 per cent.

Responsibility for Insurance Buying

The responsibility for decision-making and management of insurance is handled by an insurance department headed by a specialist in insurance, extending their services generally to companies in a group. Often the purchase department, stores department and the sales or commercial department individually look after the insurance for all movements and storage of materials that come under their responsibility. Sometimes the Financial Controller or the Chief of the Accounts Department handles all insurance by consulting the other departments, including the materials management department. Whatever may be the pattern of insurance management, the materials manager has to be intimately connected with all insurance decisions and he has also to take the responsibility for the management of insurance problems.

GENERAL INSURANCE INDUSTRY

In Britain and the US, the insurance of large industrial and commercial undertakings is handled by insurance technicians, known as Insurance Brokers. They are generally independent firms extending their services to a large number of companies. They analyse the insurance needs of their clients, negotiate with insurers both to design proper insurance covers and for settlement of claims. The broker's responsibility towards his client is absolute and he could be sued for professional negligence if the client's interests are found to be protected inadequately. But in India such services at present are not available, though we feel that in about a decade from now such insurance consultancy firms will be coming into existence. It should be possible then to transfer the responsibility of fixing the adequate and economic insurance protection to the consultancy firms. Till this happens the organisation has to keep an intelligent eye on insurance management.

The general insurance business in India was "nationalised" in May 1971. The profits in general insurance companies in the past have been fairly liberal with the result that insurance companies have generally taken a practical view of claims settlement particularly in the case of large clients whose business record from the company's point of view has been profitable over the years. Even where the scope of the cover was ambiguous or where the policy conditions were not fulfilled in their entirety, insurers were able to reach a negotiated settlement of claims without exercising their right to repudiate claims. But today the emphasis is on assisting client organisations to look after themselves so that the need for exercising discretion in claims settlement does not arise. Therefore, the need on the part of the client organisations for the greatest care in insurance management has increased.

ASPECTS OF INSURANCE MANAGEMENT

The aspects of insurance management that affect materials management are: (a) arranging an adequate insurance cover for materials in transit and storage; (b) claims management; and (c) reduction of insurance costs and the application of the concept of self-insurance.

Arranging an adequate insurance cover. We will deal with the considerations affecting the insurance of transit of incoming materials, storage, and transit of outgoing materials including finished goods.

Insurance of transit of materials is known as marine insurance and the term applies equally to ocean voyage of cargo, air consignment and inland transit by road or rail or other means of transport. The insurance policy forms used in all these cases have been standardised and the terms and conditions have also been made uniform all over the world.

Ocean voyage (Imports). The marine policy for covering ocean voyage is issued in a standard form and the terms and conditions of the standard policy are modified by a set of clauses known as Institute Cargo clauses which have also been standardised by the Institute of London Underwriters. It is not necessary for us to go into the details of the cover provided; but we will briefly touch upon some important points which have a bearing on materials management.

The marine policy is a valued policy; that is, the sum at which the cargo is to be valued is left to the discretion of the insured. The cardinal principle of insurance law is that a contract of insurance is a contract of indemnity and the law does not permit the insured to make a profit out of an insurance claim. But in the case of marine policy, it is possible for insurance to be taken for the invoice value of the cargo plus the amount of import duty and a reasonable amount of profit and incidental expenses. The principle of indemnity is thus allowed to be modified on an agreed basis. The policy specifies the cargo, the mode of transport, the voyage, the terms of the insurance cover, and identification marks for the cargo and the bill of lading details.

The marine policy covers the damage to the cargo caused by fire and perils of the sea, such as storm, heavy weather, sinking or stranding, or collision of the vessel. A marine policy for ocean voyage can be taken even after the cargo has commenced its voyage provided the insured has no knowledge that the cargo has suffered a damage. Marine policies are issued with varying levels of protection ranging from the most restricted form of cover, known as 'FPA policy', to the widest form of cover, known as 'All Risks'. Any level of protection in between these two extremes can be arranged for different premium rates. A marine policy with Institute Cargo clauses FPA attached, covers only the total loss of the cargo caused by fire and perils of the sea and partial losses are covered only where the "vessel or craft is stranded, sunk, or burnt" on the voyage. A slightly wider cover known as "WA policy" includes partial losses also arising out of perils of the sea. The "All Risks" cover provides for compensation in the event of any accidental damage to the cargo except those caused by "delay inherent or the nature of the subject matter insured".

All marine ocean voyage policies are issued subject to the transit clause which is part of all Institute Cargo clauses. The transit clause imposes important restrictions on the duration of the insurance cover and has been a subject of lot of litigation and misconceptions. The clause imposes the following limitations:

1. The insurance cover ceases as soon as the goods are delivered to the consignee or other final warehouse or place of storage at the destination named in the policy.
2. The cover ceases if the goods are stored at a warehouse which the insured elects to use for storage other than in the ordinary course of transit or for allocation or distribution.
3. The cover ceases on expiry of 60 days after completion of discharge of the goods from the overseas vessel at the final port of discharge.
4. The cover ceases on the occurrence of any one of the above events.
5. When the goods after discharge from the overseas vessel are sent to a destination other than that mentioned in the policy, the cover ceases on the commencement of transit to such destination.
6. The cover is operative only during the ordinary course of transit.
7. Where there is a change of voyage or a termination of the voyage before the goods reach the final destination, the cover is operative provided prompt notice is given to the insurer and additional premium paid.

It is a condition of the marine policy that the insured and their agents should ensure that all rights against the carriers, bailies or other third parties are properly preserved and exercised.

Steps to be taken in Insurance Management

1. The sum insured should take care of invoice value, duty, incidental expenses and reasonable profits. This point should be made clear to the insurer because occasionally policies are issued containing a wording that duty and profit are excluded from valuation.
2. Where the purchase is on CIF basis the supplier should be instructed to arrange insurance for a suitable sum. If the insurance is covered by a foreign insurer only for the invoice value, it is difficult to persuade the Indian insurer to cover duty and profit.
3. Where materials purchased on CIF basis are intended for an interior destination, the foreign supplier should be asked to arrange for insurance up to the final destination because an Indian insurer will be reluctant to issue a policy to cover transit from Indian ports to the final destination if he has not covered the ocean voyage as well.
4. The scope of the cover should be clearly spelt out. For example, where it is intended to cover leakage losses for consignment of liquid chemicals in drums, the policy should specify that leakage howsoever caused is covered under the policy. Even the all risks policy will exclude "ordinary wear and tear, ordinary leakage and breakage and inherent vice" in the absence of specific wording.
5. In cases where the value of the container is high, it is better to be clear as to whether the cost of the container has to be paid or not, even when the contents are not damaged.
6. It is possible to issue a policy to cover only duty and profit separately and for internal transit only. Such policies may be found to be helpful in fuller utilisation of import licences.
7. Where importing is a continuous process it is advantageous to take a floating policy or an open policy instead of a specific policy for each consignment.
8. Care should be taken to include war risks and SRCC risks for all ocean voyages.

Import control regulations make it unprofitable under any circumstances to arrange insurance with a foreign insurance company. It is only a drain on the foreign exchange of the country. Even where Exchange Control Regulations make it necessary for the supplier to arrange for the insurance, it will be helpful to instruct the suppliers to take a policy with an Indian insurance company operating abroad.

Inland transit. The same policy form is used for covering internal transit but the clauses attached to the policy are different and the scope of the cover is also substantially different. The following are the important features of the inland transit policy:

1. The terms of the cover should be as wide as an "all risks" policy or as restricted as "road risks" or "rail risks" policy.
2. The "all risks" cover is subject to the same limitations as mentioned earlier. The 'road risks" or the "rail risks" policies cover the risk of loss or damage due to fire, lightning, collision, overturning or accidents of a like nature while being conveyed in the vehicle and during transhipment prior to reaching the destination. The policy covers the risk of fire during storage. It is possible to add specified risks to a "road risk" policy instead of taking the costlier "all risks" cover.
3. War risks on land cannot be covered by insurance companies. But riot and strike cover is an exclusion under the standard policy which should be included specifically by payment of additional premium.
4. All inland transit policies restrict the duration of the cover in a way which is substantially different from the limitations applicable to ocean voyage. Here, the insurance commences with the loading of the material into the vehicle, extends during the transit including transhipment, prior to reaching the destination and ceases seven days after the arrival of the vehicle at the destination town or on delivery, whichever is earlier. Where the cover is extended specifically by the standard 'Warehouse to Warehouse clause', it continues till the goods reach the final warehouse, subject to a limitation of seven days after reaching the destination to town. During the course of the storage at destination, only fire risks are covered.

The extent of insurance cover to be taken for incoming materials and for finished goods should first be determined, by asking questions similar to the following:

1. Should the insurance be done on restricted "road risks" or "rail risks" basis or on "wider" terms?
2. In the case of incoming materials, should the insurance be left to the supplier stipulating that the suppliers should replace the damaged materials?
3. In the case of transit of finished goods, should the insurance be on restricted basis, wider basis or on self-insurance basis?
4. If it is to be self-insurance, what are the risk accumulation points and what is the extent of risk accumulation and what is the maximum possible loss arising out of an accident?
5. If insurance is arranged, what is the period of storage at the destination for which the cover has to be extended?

The above questions have to be answered in the light of the following conditions prevailing in our country. Where transport carriers, both Railways and lorry transport companies, are liable for compensating the consignor for negligent handling of goods, the time and effort required to recover from them is substantial and their liability in the case of some catastrophic losses like riot damages is doubtful. Further, lorry transport companies in India at present are not financially sound enough to compensate their clients in the event of a fairly big loss. The extent of insurance protection they have to save themselves from liabilities, is also negligible.

Insurance of Stores

This presents relatively minor problems. During storage, goods are generally covered against fire, explosion, riot and strike, flood and earthquake, under a fire policy, and occasionally for burglary and theft risks under a burglary policy. The following are some important considerations which influence insurance decisions:

1. Where the stock is huge, insurance for burglary and theft risks is not sought for.
2. A fidelity guarantee policy for the key personnel controlling stores often turns out to be a wise investment since the cost for such insurance is relatively low.
3. Where the stocks fluctuate in value during the course of a year, the insurer can fix up a declaration policy for the fire and allied risks, under which premium is charged only on the average value of the stocks.
4. The fire policy and all property insurance policies prohibit under insurance by the incorporation of the Average Clause, under which the insured is treated as his own insurer to the extent of under-insurance and only a portion of the claim depending on the extent of under-insurance is recoverable from the insurers.
5. The premium rate for warehouses is dependent on the nature of storage in each non-communicating part of the building and therefore it is possible to economise in premium amount if you can ensure that hazardous and non-hazardous materials are stored separately and storage warehouses do not communicate with the manufacturing section.
6. Where new machinery is purchased it is advisable to cover the risk of transit, storage and erection under a single policy known as 'marine-cum-erection policy'.

Claims Management

Negotiation and correspondence on claims and delay in settlement of claims could be as costly in the long run as outright repudiation of claims. To achieve quick and full settlement of claims, proper care should be taken at the time of taking insurance to see that the risks are covered adequately and unambiguously on the lines discussed above. Further, all policy conditions during the policy period and after the claim had arisen must be fulfilled. For this, the following are the important points to be considered:

1. Wherever there is a chance of the insurance cover terminating before the materials reach the ultimate destination, written intimation to the insurer seeking extension of the cover should be given.
2. Whenever there is a change in the ordinary course of the transit, written intimation to the insurer would save a lot of trouble.
3. As stated earlier, it is a condition of a marine policy that the insured and their agents should preserve their rights of recovery against the carriers and bailees and transfer those rights to the insurer. Where this is not ensured the claim against the insurer is prejudiced.
4. The insured or their agents must claim on the carriers and on the Port Trust authorities for any missing packages according to regulations; and where they do appear to be damaged, a survey in the Docks must be conducted by a carrier's representative and a claim should be lodged on the carriers. In the case of rail and road transport open delivery certificates must be demanded and detained in the case of an apparent damage to the consignment.
5. In no circumstances should clean receipts be given where goods are in doubtful condition except under written protest.
6. Notice in writing must be given to the carrier's representative within three days of delivery if the loss of damage was not apparent at the time of taking delivery.
7. It is also essential that prompt written intimation of damage giving rise to a claim should be given to the insurer and a survey arranged. An insurer considers a claim properly lodged and demanding their serious attention only when the documents such as original policy, bill of lading or railway receipt or lorry receipt, invoice, submission of relevant documents concerning the claim, survey report, claim bill, letter of subrogation are received by them.

Minimising Insurance Cost and Self-Insurance

There are only two ways of minimising expenditure on insurance: First, by reducing the element of risk involved in materials movements, and, secondly, by introducing self-insurance to take care of small and fairly predictable losses and seeking the insurer's assistance only for protection against catastrophic losses.

Risk Improvement and Minimisation of Loss

The nature of packing, the mode of transport and the extent of care exercised in materials movement have a bearing on the probability of losses occurring and on the quantum of such losses. Whenever there is an improvement in any of these, the insurers should be persuaded to consider a reduction in premium rates. It is sometimes profitable to discuss risk and loss minimisation measures with the insurers, themselves, who may come out with suggestions based on their experience over a wider range of clients.

It appears a sound policy, from the point of view of cost reduction, to consider self-insurance for certain risks, especially where the size of operations is large. The cost of processing claims has to be taken into account while calculating the expenditure on insurance. Self-insurance could operate more profitably for the transit of finished goods than for incoming materials, because, in the case of finished goods, all the risk factors attendant on transit are within the organisation's control and the experience of losses over the past few years should be readily available to ascertain the pattern of transit losses. But to transfer the insurance responsibility to the carriers is an unwise move. Self-insurance could be introduced progressively by first excluding claims for less than a specified amount and working up to a stage where only the most restricted form of insurance to take care of catastrophic losses is taken. But these are decisions which have to be taken in each case depending on the circumstances, taking into account the accumulation of risk, the maximum possible loss arising out to an accident, the administrative expenses in dealing with insurers, suppliers and buyers, regarding claims and the tax angle.

A systematic and periodic, preferably, an annual review of the existing insurance arrangements of the organisation, the cost and benefits, should be undertaken. The insurance needs of a large and complex organisation are ever changing and the insurance cover designed with care today may not remain a perfect fit two years from now. Exhibit 19.1 gives the road risk of an insurance company.

EXHIBIT 19.1
ROAD RISK CLAUSE

1. This insurance commences with the loading of each bale or package into the vehicle and covers the risk of loss or damage occasioned by fire, lightning, collision, overturning or accidents of a like nature whilst being conveyed in the vehicle and during trans-shipment prior to reaching the destination town. Risk to cease seven days after the first arrival of the vehicle at the destination town mentioned in the policy, or on delivery whichever shall first occur.
2. Warranted that no liability attaches hereunder, in the event of the property insured hereby being transported by lorries propelled by producer gas.
3. Warranted that no liability shall attach to the Company under this policy in respect of goods lost or damaged due to a peril insured against, whilst in the custody of the carrier, unless copy of a claim, provisional or otherwise, lodged on the Carrier by the Insured, accompanies claim documents.

20

Buyer-Seller Relationship and Ethics

NEED FOR CORDIAL RELATIONS

The buyer as well as the seller know very well that good relations are essential for the smooth running of the business. Yet there are many suppliers who frequently fail to meet the delivery schedules and fail in quality standards. Vendors on the other hand complain about delayed payment, cancellation of orders and increased rejection due to rigid inspection. In any organisation, it is the prime responsibility of the materials manager to keep relations smooth.

The buyer and seller must recognise that they are partners in progress and as such, good relationships between them can go a long way in adding to the vitality of on-going professional relationships as well as in increasing the effectiveness. The underlying philosophy should be based on mutual confidence and should be one of inter-dependence. The primary objective of the relationship is to buy and sell for a profit. This should be recognised by the buyer as well as the seller. The supplier can assist the buyer through ideas for cost reduction, new products and service even during a crisis. Such healthy relationships are possible only when both the parties are competent in their profession and respect the principles of human relations. For example, if the supplier lacks the know-how and facilities to meet the specifications sought by the buyer, he will be of little help. Similarly, if the buyer lacks techniques of negotiating with the suppliers, he can never hope to develop a meaningful relationship with the supplier. The buyer should consider a good vendor as an intangible asset of his organisation.

CONTROL IN PRACTISING RECIPROCITY

Reciprocity is the practice of selecting one's own customers as suppliers. In fact, under reciprocity, the materials manager may even accept slightly higher prices from a supplier because he is also a customer. Sometimes this practice may result in strained relations between other suppliers and the buyer. Whereas reciprocal buying is advantageous in boosting the sales and fosters goodwill with the suppliers in question, it is not a desirable practice as in many cases such suppliers are very small in number. This will be especially so in the case of companies which have only one product to offer when many of the suppliers will find no use for the product.

BACKDOOR SELLING

Suppliers' salesmen form an excellent source to developing healthy and effective relations. Many companies give a "welcome booklet" to the salesmen giving details, such as names and locations of the buyer responsible for different commodity classifications, a brief note on organisational arrangements, physical locations of purchase departments, parking lots, etc. Also, the booklet will give norms of the company with regard to suppliers' salesmen. They will be required to use only the "front door", i.e., they should have dealings only with the concerned buyer. On no occasion, except with the buyer's explicit approval, should they meet the operating personnel. Several arguments can be put forward to justify this. Firstly, it is the buyer's responsibility to meet the seller's salesmen—for which he gets paid. Secondly, salesmen will unnecessarily waste the time of operating personnel in discussing problems which are better sorted out by the buyer. Further, the user should not be directly approached by the vendor for ethical considerations.

FRUITFUL AREAS FOR IMPROVEMENT

Buyers should make it a point to look upon their relationship with suppliers as one of inter-dependence. It is essential that buyers should understand the needs of suppliers. Suppliers expect fair and honest dealings, ethical practices and a minimum amount of rush orders. During negotiations they prefer to deal with buyers who have sufficient authority, and who are reasonable and well informed. The materials manager ensures that personal contacts are encouraged. In this context, the buyer should periodically visit the suppliers' plant and assess the vendor's strengths and weaknesses.

Many buyers have found it useful to organise meetings of their suppliers periodically, to develop personal contacts and to acknowledge inter-dependence. Some buyers believe in organising 'informal meetings' with major suppliers to explore possibilities for cost-reduction. The following steps will ensure that buyer-seller relationships are kept at a healthy level:

1. Inform the vendor of the major purchase policies. This will save his time and effort.
2. Make it a point to give the vendor's salesmen a courteous reception when they visit the plant and interview them as per pre-arranged schedule. It is essential to conserve the time and give others an idea of how long they should wait.
3. Give adequate notice to the vendors so that they can be ready with quotations. Recognise any extra service rendered by vendor's salesmen through a letter of appreciation.

4. Be ready with "home work" so that negotiations can be businesslike and to the point. This requires the obtaining of complete quotations, tabulations and vendor rating information. When a buyer is armed with such information, his negotiations with the supplier are purposive and the supplier also saves his time and effort.
5. Keep rush orders at a minimum and inform the unsuccessful bidders as to why they have not received the orders.
6. Allow permissible tolerances in inspection and forward inspection reports to the vendors with suggestions so that they could improve.
7. Ensure the removal of rejected materials from the buyer's premises promptly.
8. Do not cancel orders with suppliers at the last minute. Give sufficient notice to the supplier for the cancellation of an order.
9. Clear the outstanding bills of the suppliers as expeditiously as possible.

Many other factors, such as fair dealing, living up to the agreements in both letter and spirit, assisting suppliers in development, proper appreciation of each other's difficulties and mutual confidence can help in moulding cordial buyer-seller relationship. It is necessary that the buyer should have an adequate and competent staff who could assist him in scheduling interviews with the suppliers, preparing the necessary groundwork, soliciting and refining the opinions of the suppliers for corrective measures and so on.

Attempts must be made by the materials manager to evaluate his department's relations with suppliers from time to time. He should compare, over a time period, factors such as the number of visits made to suppliers, plants, number of meetings held with the suppliers to sort out mutual difficulties and complaints regarding gifts, backdoor selling, etc. The materials manager should accept the philosophy of interdependence with suppliers and try to develop a climate of mutual trust and confidence.

IMPORTANCE OF PUBLIC RELATIONS

We will now discuss how good public relations and sound ethical principles contribute towards better buyer-seller relationship. Public relations is very important for the image of an organisation and significantly contributes to the buyer/seller relationships. The Materials Manager, by virtue of his constant interaction with the environment and vendors, obviously is in a position to contribute significantly to the public relations efforts of the company. It is quite possible that the materials manager in discharging his duties may have good relations with some and not-so-good relations with others. The underlying philosophy in any public relations effort is the ability of the members of the organisation to put themselves in the position of the persons with whom they deal and view the situation from their angle as well. In other words, for healthy public relations the members should have sympathy.

Public relations may be looked upon as an effort to develop a favourable climate and attitude towards the company or to carve a good image for the company. It can also be defined as an effort through which the company hopes to ensure the understanding and goodwill of its customers, employees, suppliers, as well as that of the public in general. By this definition, it follows that this effort consists of two parts. The first part concerns internal relations, and the second pertains to external relations.

INTERNAL PUBLIC RELATIONS

We have seen that the materials management function, in order to be successful, will have to work with other functional departments. Naturally, an excellent rapport will ensure that materials management can function very effectively. This rapport can be achieved through the realisation of the interdependent nature of the relationships, without tending to overemphasise one function in relation to others.

There are very simple yet effective ways in which such a rapport between materials management and other departments can be established. Apart from formal communication channels, every organisation has its *informal* channels. These channels will be of great help in getting to know the other departments' problems with materials management so that remedial measures can be taken. There are several reports which are of interest to user departments. For example, a report on substitute materials, transport bottlenecks, dealy in supply position, incoming materials quality, etc., is of great assistance to the production department. Similarly, reports on price trends may be of interest to the finance department. Such formal reports in combination with informal channels can considerably foster good relations within the company.

EXTERNAL PUBLIC RELATIONS

Normally, the materials manager is the first point of contact between his organisation and the suppliers. In his day-to-day activities he gets ample opportunities to project a forceful, effective PR image. This he can achieve only by doing a sincere job in each of his interviews with various persons. He must recognise the value of time, show understanding and appreciation of the services, information given and suggestions made by the representatives of the suppliers. It is likely that with some representatives no business is possible for a long time. Yet a diplomatic knack of tackling them will keep future business open to both the parties.

During the course of his job, the materials manager writes to many outside agencies. This could act as an effective public relations tool. Prompt replies suggestions, quick intimation about price changes, courtesy in written communication and telephonic conversation, prompt reception to salesmen, courteous interviews, etc., can go a long way in building up the image of the company, in general, and the materials management department in particular.

Quite a number of organisations have found it desirable to periodically feed the suppliers about the quality and delivery performance and furnish inspection reports, thus pointing out the areas where improvements are desired. Such initiatives really strengthen the rapport between the suppliers and buyer. We have also noticed that plant visits are of significance in this area.

TRAINING AND PUBLIC RELATIONS

In the backdrop of fast changing technology and an ever growing economy, the function of the materials manager is becoming quite challenging. A seller's market as well as the increasing impact of government policies together make the challenge. What is more, top management is increasingly expecting the materials department to play a vital role in the company's cost reduction aspects. Therefore, additional skills have to be imparted to prepare an executive to react effectively to changes. An exposure of the materials manager to sensitivity training and workshops can prepare him to function effectively within the organisation as well as with outside agencies and help him to project a favourable image. Role playing and case studies are some of the well known training techniques which will prepare the materials manager to handle public relations functions effectively.

ETHICS IN BUYER-SELLER RELATIONSHIP

The contribution of the ethics to buyer-seller relationships has been discussed in materials management literature at great length by several authors.

Ethics, though closely related to personal values, are obviously influenced by moral standards prevalent at the corporate level, industry level, national and international levels. Ethics have more significance to a materials manager as he is dealing with many suppliers, evaluating them and awarding them contracts. In India, materials managers of some big organisations are responsible for spending money to the tune of Rs 100 crores per year. Attempts will naturally be made to curry favour with him and as such the materials manager is subject to temptations day in and day out. Therefore, it is very necessary for the materials manager to resist such temptations and be above board. Ethical practices can be fostered by devising procedures and systems for control purposes as well as by selecting buyers who have unquestionable integrity, character and high moral standards.

Policy decisions pertaining to practice of reciprocity, personal buying, acceptance of gifts, and entertainments can straightaway solve half the problems. Supposing the management policy is not to accept gifts valued more than Rs 20 this will automatically curtail conspicuous "gifts" from suppliers. Besides being honest and refusing to accept favours from suppliers, an ethical buyer should also be equitable in his dealings with suppliers. That is to say that he should not be influenced by personalities and should be guided solely by objective considerations. Also, during the process of negotiation the buyer should not resort to unethical practices. In practice, many organisations have developed a code of ethics which the buyers are expected to practise in their profession. It is rather unfortunate that organisations adopt double standards—one for the acceptance of gifts by the buyers and another for their salesmen using gifts for sales promotion!

The code of ethics formulated by National Association of Purchasing Agents,

USA, is given below. In India, too, a code of ethics has been developed on the same lines.

PRINCIPLES AND STANDARDS OF PURCHASING PRACTICE

1. To consider, first, the interest of his company in all transactions and to carry out and believe in its established policies.
2. To be receptive to competent counsel from his colleagues and to be guided by such counsel without impairing the dignity and responsibility of his office.
3. To buy without prejudice, seeking to obtain the maximum ultimate value for each dollar of expenditure.
4. To strive consistently for knowledge of the materials and processes of manufacture and to establish practical methods for the conduct of his office.
5. To subscribe and work for honesty and truth in buying and selling and to denounce all forms and manifestations of commercial bribery.
6. To accord a prompt and courteous reception, so far as conditions will permit, to all who call upon a legitimate business mission.
7. To respect his obligations and to require that obligations to him and to his concern be respected, consistent with good business practice.
8. To avoid sharp practice.
9. To counsel and assist fellow purchasing agents in the performance of their duties, whenever occasion permits.
10. To cooperate with all organisations and individuals engaged in activities designed to enhance the development and standard of purchasing.

Apart from this, the materials manager must adhere to some other norms. He should not encourage any supplier to get the quotations of others. He must treat them as confidential. He should not misquote nor should he try to give false promises and hopes. He should not use his buying power to force small suppliers to sell at uneconomical costs. He should try and avoid awarding contracts to his relatives and must never have a financial interest in supplier's firms. In other words, he should not have a conflict of interest while dealing with outside agencies, and should set an example for the others.

PROFESSIONAL AGENCIES AND TRAINING

One of the hallmarks of a professional person is that his devotion to excellence in his chosen field outweighs the selfish desire for personal gains. In this context, agencies like the National Association of Materials Management, the Materials Management Officers' Association of India and the Indian Society for Materials Management can play a significant role in promoting the image of the materials management profession. (It is heartening to note that attempts are being made to form an All India Association out of the above agencies.) The advantages of membership in such agencies are not far to seek. The members get an opportunity to meet and exchange ideas. Such agencies do bring out periodicals and the members can effectively use this media to air their views and contribute professional articles. The suggestions of these agencies are considered by the government while formulating import policies or in allocating critical raw materials. These professional agencies can aid the individual materials managers in improving buyer-seller relationship as well as the ethical standards of the profession.

The professional agencies can also play a useful role in formulating detailed syllabi for training programmes at different levels with the cooperation of institutions like the Administrative Staff College of India. The need for training in the materials management field has been emphasised by various reports. For instance, the fortieth report of the Committee on Public Undertakings has emphasised this aspect as follows: "There is a general feeling that materials management does not require any specialised knowledge. It is perhaps for this reason that there are practically no regular arrangements in the country for training of personnel at various levels, for appointment to materials management posts. The efficiency of the materials management department will depend on the qualifications and experience of the personnel appointed therein. It is therefore essential to arrange regular institutional training in materials management and for exchange of ideas and information through seminars, etc."*

***Fortieth Report on Materials Management in Public Undertakings* by Committee on Public Undertakings, Lok Sabha Secretariat, New Delhi, p. 62.

Part III
WAREHOUSING AND STORES MANAGEMENT

21

Stores Management

PURPOSE OF STORES

Stores play a vital role in the operations of a company. It is in direct touch with the user departments in its day-to-day activities. The most important purpose served by the stores is to provide uninterrupted service to the manufacturing divisions. Further, stores is often equated directly with money, as money is locked up on the stores. The functions of stores can be classified as follows:

1. To receive raw materials, components, tools, equipments and other items and account for them.
2. To provide adequate and proper storage and preservation to the various items.
3. To meet the demands of the consuming departments by proper issues and account for the consumption.
4. To minimise obsolescence, surplus and scrap through proper codification, preservation and handling.
5. To highlight stock accumulation, discrepancies and abnormal consumption and effect control measures.
6. To ensure good housekeeping so that material handling, material preservation, stocking, receipt and issue can be done adequately.
7. To assist in verification and provide supporting information for effective purchase action.

In India, owing to stiff supply positions, 4 to 6 months' inventories are not uncommon and, in fact, for certain imported items, it could be as high as 24 months' stock. In this context, stores management assumes greater importance.

LOCATION AND LAYOUT

More often than not, in the matter of locating the stores, materials management is rarely consulted. The normal practice is to locate the stores near the consuming departments. This minimises handling and ensures timely despatch. In stores layout, the governing criteria are easy movement of materials, good housekeeping, sufficient space for men and material handling equipments, optimum utilisation of storage space, judicious use of storage equipments, such as shelves, racks, pallets and proper preservation from rain, light and other such elements. These problems are more important in the case of items that have a limited shelf life. Other important factors governing the location are the number of end users and their location, the volume and the variety of goods to be handled, the location of the central receiving section and accessibility to modes of transportation such as rail or road.

Since stores have to be nearest to the user, large organisations usually have stores attached to each consuming department, whereas receiving is done centrally. Items of common usage are stocked in the Central Stores so that inventory is kept at an optimum level. These factors are considered at the planning level of layout.

In the case of warehouses stocking finished goods, factors such as proximity to ports, railway lines, quality of roads, availability of power, etc., become quite important. It is also important that the stores are constructed with a futuristic orientation, so that sufficient flexibility for expansion needs is inbuilt. The activities of receiving the goods, stocking in appropriate locations, material handling and issues must be done swiftly and economically. The stores building must have adequate facilities for preservation of stores. Sometimes facilities, such as cold storage, heating equipments, airconditioning and similar facilities may be required. These should be planned in advance. Comfortable working conditions must be provided to the stores personnel to get maximum efficiency and morale. The important factors in the design of stores building can be summarised as follows:

1. *Lighting.* Clear and adequate lighting is a must for a proper work environment. Lighting effects can be accentuated through a judicious choice of colours for the walls. For stores personnel who work day in and day out in the stores receiving, checking, stocking, handling and issuing goods, a pleasing environment goes a long way in reducing monotony. Any attempt to reduce these facilities will prove to be false economising in the long run.

2. *Safety.* This factor is perhaps the most important aspect. In stores a large volume of goods are handled every day. Accidents considerably reduce the morale and effectiveness of the system. The following measures are necessary if accidents are to be checked:

(a) Safety consciousness should be instilled in the minds of stores personnel through training programmes, visual aids and literature.

(b) Safety appliances, such as goggles, hand gloves, etc., must be provided and their use must be encouraged.

(c) Good housekeeping is essential. This means that gangways must be clean, adequately wide so that movement of forklifts, trolleys and industrial tractors is smooth. Stocking must be in appropriate locations so that handling is minimum.

(d) All stores equipment must be kept in good order. This includes adequate maintenance practices with regard to forklifts, overhead cranes, trolleys, conveyors, etc. Operators must be trained in safety aspects so that safety precautions are not overlooked.

(e) Healthy competition can be stimulated by installing 'safety awards' and cash prizes which bring recognition to the concerned stores personnel for safety practices. This also motivates others to practise safety.

(f) Provision of fire fighting facilities is necessary especially where inflammable materials are stored and handled. In point of fact, large organisations have a well-maintained fire fighting equipment with the stores in preparedness. This has in the long run reduced losses and reduced insurance expenses. Fire extinguishers, fire escapes, alarms and sprinklers must be available and personnel should be familiar in handling them.

(g) Other factors which merit attention include provision of toilets, routine maintenance equipments, safe electrical wirings, etc.

COST ASPECTS AND PRODUCTIVITY

It is erroneously averred that every cubic metre of space must be utilised by stocks for high efficiency. Very often such stocking may drastically cut down the speed of material movement and create bottlenecks apart from affecting overall safety. Therefore manoeuvring needs for handling equipments and for minimising the time required for receipt, stocking and issue, must be borne in mind at the planning stage in order to ensure real efficiency.

Costs involved in stores can be analysed under two heads, vis., fixed and variable. Fixed costs are to be incurred irrespective of the utilisation of the space of the stores. They include money spent on land and buildings, rent, interest, repairs, maintenance, insurance, etc.

Variable costs vary with the volume of throughput. They consist of handling costs, damages, deterioration, obsolescence, etc. Obviously, when the throughput or the volume of goods handled is high, the total cost per tonne is low. This should be the aim of the Stores Manager in order to optimise the costs in stores.

PROBLEMS AND DEVELOPMENTS

It is an unfortunate fact that Stores Management has been regarded as a clerical function. In the gamut of materials management, stores is considered as the least glamorous and it never attracts talent. It is forgotten that the Stores Manager is probably the custodian of the single largest group of current assets and plays a pivotal role in ensuring smooth production, besides assisting purchase activities through timely support. This is the major problem and challenge that faces the stores manager today. Many decisions is stores management, such as selection of racks, bins, handling equipment, safety practices, codification, training personnel and accounting, call for considerable skill and an ability to coordinate with other departments as well as with outside agencies. These aspects should be highlighted and appreciated so that the stores function is given due importance. Other areas in stores, such as records keeping, movement analysis to reduce obsolescence, surplus and damage are critical to the profitable operation of the firm and the stores manager faces a challenge in these areas as well. In many organisations, the scrap yard also comes under the control of the stores manager. This is an entirely new responsibility calling for the ability to maximise returns on the disposal of scrap. The Chief Stores Officer has under him separate officers for the functions of receipt, issue, kardex and sub-stores. Besides coming into contact with the production, purchase, maintenance, inspection and finance departments within his organisation, he has to come into contact with outsiders like suppliers, transport carriers and bankers. In order to meet such challenges, the importance of the stores function should gradually gain momentum and qualified engineers should be posted as Chief Stores Officers reporting to the materials manager.

NEW DEVELOPMENTS IN STORING

We will briefly discuss a new development in storage practice in India. This is known as dry storing. This is a fairly new concept in sack/bulk stores and has resulted in an extremely versatile and economical method of dry storage. Available initially in four sizes, that is 250, 500, 750, and 1,000 tons; the stores are suitable for storing any type of bagged material, as well as grain and all other types of dry foodstuffs and materials. Made from thick rotproof butyl rubber laminate and heavy duty steel mesh, they are completely portable, allowing their use to be either temporary or permanent and they can be moved from one site to another in a matter of hours. An outstanding advantage is simplicity of operation, and no mechanical loading aids need be used unless desired.

The stores are circular in shape, with an outer support wall made up to 7' 10" high 5 gauge weldmesh. To the inside of this is secured a thick lining of butyl rubber laminate, which consists of 0.040" layer of black butyl rubber faced with a 0.010" thick skin of white EPDM rubber bonded to it. This lining passes up and over the top of the side-wall, to which it is secured by a continuous clip/rim, and a 0.050" thick butyl rubber floor is bonded to the sidewalls.

The effect is that of a giant bin, its diameter according to the specific model varying between 38' 0" and 68' 0". The work/storage area inside the bin is not obstructed in any way, and the roof/lid fitting over it, again made from thick butyl rubber laminate, is intended to be supported at the centre by the sacks actually being stored. At the sides, it is clipped to the rim of the sidewalls by clamps, which are tensioned by a simple strainer device and not only hold the lid firmly in place but also effectively seal the inside off from atmosphere.

Sacks can be loaded into the store by various methods. Thus, a stairway can be built up to the sidewall so that the sacks can be quickly passed into the store or a lorry can be backed up to the sidewall and sacks passed over from the vehicle tailboard. A third alternative is the use of a mechanical conveying system. For bulk materials the store can be loaded manually or mechanically. Once inside the store, sacks are loaded from the centre outwards, higher at the middle up to approximately 15' 0" in the largest model and tapering towards the sides, the lid being conical in shape for this purpose. Because the storage is both waterproof and airtight, bagged grain and other products can be safely stored for long periods.

All that is necessary for installation is a suitable area of levelled ground with a coating of four inches of sand on top. If the store has to be moved, the steel mesh sidewall can be unbolted and rolled up, and the butyl rubber envelope of the store complete with lid, folded into a compact package.

New developments have been made on the field of preservation of stores and prevention of pilferage. A detailed account of various preservation methods for different categories of items such as oil, paints, rubber goods, timber, etc., is given in the book entitled *Materials Management* by Deb*. For example, electrodes must be stacked in dry places. Bearings must be properly greased. Wood stores must be free from moisture and so on. Industries have taken several measures to cut down losses through pilferage. Large organisations have found it useful to maintain intelligence squads, security personnel and electronic alarms. Scrap yards where "jet age" scrap such as tungsten, copper, etc., are dumped are even guarded by dog squads. Frequent and surprise checks are made by top management and strict penalising action is usually taken on those found guilty so that it may serve as a deterrent to the others.

Tool stores. Special reference has to be made with regard to tools stores. By and large the tools stores items are very costly and the usage points are scattered all over the plant. These two factors have made it essential to develop a unique stores management concept. Generally, there is a central stores where the tools are required from outside suppliers and tool room. Here, accounting is done with regard to quantity received and quantity accepted after inspection and tests. Then the tools are sent to decentralised locations known as tool cribs. Tool cribs are located near the usage points. For example milling cutters, arbours and related tools may be kept near the gang milling line. Here, the issue is controlled on an imprest basis. Usually workers are required to produce scrapped tools before they are issued with fresh tools. It is quite common to find tool stores equipped with regrinding facilities so as to maximise the tool life and at the same time ensure quick availability.

*A. Deb, *Materials Management*, Academic Publishers, Calcutta, 1969.

22

Stores Systems and Procedures

Broadly the systems in stores can be studied under three areas, namely, receipt, stocking and issue. The flow chart in Exhibit 22.1 will give a good idea as to the activities involved and the responsibilities in the stores function. It will be seen that at every stage a great deal of information is required for checking, controlling and feedback purposes. Well designed stores systems and procedures ensure timely information for decision-making, particularly because stores is the starting point of all activities for control. Let us briefly consider the systems and procedures in each area.

RECEIPT SYSTEM

This can be divided into receipts from outside suppliers and receipts from internal divisions. Systems for receipt start even before the time when the material actually reaches the plant. When a purchase order is placed, a copy is sent to the stores, indicating quantity and delivery date. These should be arranged in a chronological sequence so that the stores manager can at any time estimate the volume of receipt. This also helps in planning labour contracts when unloading activities exceed a particular limit. This is the first step in the stores system. Secondly, suppliers, once they despatch the goods, normally send an advice note to the stores. This provides information on the date of despatch, carrier details, description of the consignment and value. This is sent in advance so that quick and easy clearance may be done. The third stage is the document prepared by the transport carrier. Very often different suppliers employ different transport organisations for transportation of their materials. These transport organisations usually send consignment notes to the stores concerned. Railways and private transport organisations are examples in this respect. These three documents, namely, copy of the purchase order, supplier's advice document and the consignment note, enable the Stores Manager to organise and plan for expeditious clearance of materials and minimise costly demurrages. In some cases, suppliers send a packing slip detailing the contents in the package.

PHYSICAL SYSTEMS

When the anticipated day of delivery comes, the above documents are tallied for

EXHIBIT 22.1

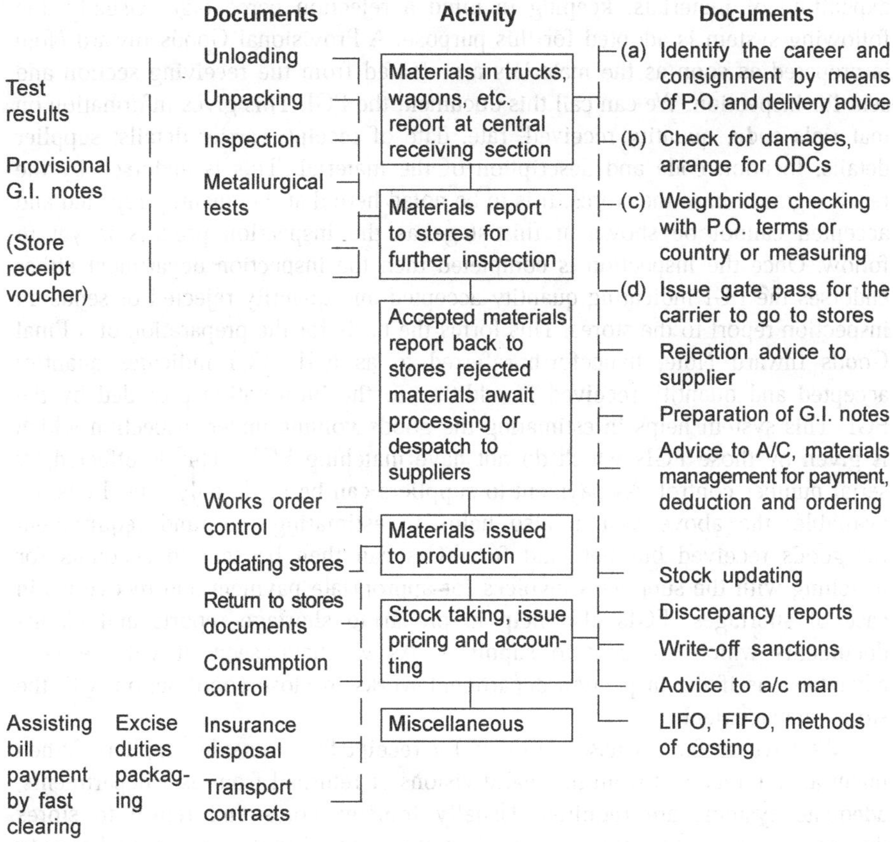

identity of figures with respect to quantity and value. When the consignment arrives it is identified with the help of these documents. Then it is physically verified using weighbridges, measuring devices, tapes, etc. When the volumes of receipts are high, this process could be unwieldy and may prove to be a bottleneck. In such cases arrangements are made for inscribing the tare-weight of the trucks and wagons. This reduces frequent weighing. When there is no significant difference between actual amount and the amount shown in the three documents, the consignment is sent for inspection. If shortage is observed then it calls for additional procedures. The time element is very important as shortage claims will not be honoured when they are time-barred. Therefore, the documents are prepared and indexed datewise detailing the quantum of shortage and value. This may require the stores personnel to take open delivery in the presence of the transport organisation's official and get the shortage endorsed.

Simple systems, such as those discussed earlier, could result in savings through less demurrages and higher returns on claims. Once the consignment is cleared at this stage, then the inspection stage follows. It is known that at any time a good amount of materials await and/or undergo inspection and as such are not available readily for issue to production. Also, the materials manager needs to know the amount of materials existing at the inspection stage, so that he can schedule his expediting of materials, keeping in mind a rejection percentage. Usually, the following system is adopted for this purpose. A Provisional Goods Inward Note is prepared as soon as the materials are cleared from the receiving section and sent for inspection. We can call this document the PGI. This gives information on materials code, quantity received, rate, date of receipt, carrier details, supplier details, location code and description of the material. This is endorsed by the receiving section of the stores. It is to be noted here that the quantity rejected and accepted cannot be shown at this stage as the inspection process is yet to follow. Once the inspection is completed then the inspection department either endorses the PGI indicating quantity accepted and quantity rejected or sends an inspection report to the stores. This forms the basis for the preparation of a Final Goods Inward Note, henceforth referred to as FGI—FGI indicates quantity accepted and quantity received in addition to the information provided by the PGI. This system helps in estimating the goods volume under inspection which is given by those FGIs which do not have matching PGIs. This is effected by serial number control. As payment to suppliers can be made only after FGIs are available, the above system also helps in estimating the fund requirement for goods received but not paid for. FGIs can then be sent to accounts for matching with the supplier's invoices for appropriate payment and recoveries in case of shortages. FGIs also help in preparing shortage reports and claims documents which are sent to suppliers. As we have seen, it will be very advantageous if the inspection department works in close coordination with the stores management.

We have so far discussed systems for receipt from outside suppliers. When materials are received from internal divisions or returned from user departments, adequate systems are required. Usually transfer notes and return to stores documents are used for this purpose. Sometimes the scrap is also handed over to the Stores Department. Scrap cards indicating the nature and weight of the scrap are prepared for such purposes. Examples of G.I. Notes and discrepancy notes are shown in Exhibits 22.2 and 22.3.

We will discuss the various organisational arrangements in inspection. Some organisations have found it advantageous to integrate inspection department under materials management. This reduces paper work, results in lesser materials movement and swift clearance of documents leading to prompt payment of bills and timely claims. However, there is a conflict of interests between quality and service to the user department and this is the major drawback in this arrangement.

In a few organisations inspection is integrated under the respective user departments. Here also, when production targets are ambitious, a tradeoff is likely to occur in quality which ultimately affects the company image.

Inspection in many organisations is an independent activity where the manager in charge of quality control is responsible only to the General Manager. This arrangement ensures protection of quality standards, although at times this may lead to overemphasis on not so important quality issues and result in a draining of the company's resources. Detailed aspects of inspection are discussed in the next chapter.

EXHIBIT 22.2
FINAL G.I. NOTE

Material Part No.: Serial Number:
Description:
P.O. Reference:

Carrier details	Supplier details	Inspection report
Truck/Wagon R/R: Consignment note	Supplier code Supplier name	Test results Conclusions

Quantity received	______________	Damage/Shortage ______
Quantity accepted	______________	Shortage claim ref. _____
Quantity rejected	______________	

Sd/- Central Receiving Section Date:	Sd/- Inspection Department Date:	Sd/- Stores Department Date:

Copies to:		
	Accounts	: Attention bills/Insurance and claims
	Indentor	: For information
	Materials	: For updating and expediting
	Management Dept. Stores	: For stock records and reference

STORING PRACTICES

At the end of the receipt and inspection stage, stocking follows. This is the most under-rated function in stores management. Stocking involves routine activities like sorting out materials coming at the end of inspection process and storing them in their locations. In big organisations, the volume and variety of materials to be stocked are so high that areas within the stores are demarcated as follows:

1. Area where materials are to be stocked for inspection (This is usually earmarked near testing laboratories and inspection outfits);
2. Area where materials coming out of inspection and accepted for use within the plant are to be stocked; and
3. Area where materials rejected are to be stocked so that they could be despatched to the suppliers concerned. (This is located near wood working section and sidings for suitable packing and onward despatch.)

Stocking is very important for easy location, proper identification, and speedy issue to the consuming department. This process is very crucial in warehouses where thousands of parts are stocked for meeting consumer needs.

Another important aspect is the need to specially stock excisable items. Certain items are subject to inspection by government authorities before issue to consuming departments. For this purpose bonded stores are used. This is nothing but a special store within the main stores enabling easy identification of such items.

EXHIBIT 22.3

From	To
XYZ Co. Ltd.	Carrier
	Consigner

Discrepancy Report

Consignment note ref.: Invoice No.

Date:

Dear Sirs,

We have received..............packages against the above Consignment Note on..............The consignment was externally in good condition. The consignment was opened on ....and we have found the following discrepancies:

Description	*Quantity invoiced*	*Quantity received*	*Shortage*	*Excess*

1. Please arrange to make good the shortage/raise invoice for the excess.
2. We have deducted the value from the invoice for the shortage.
3. We have arranged to return the excess and shall adjust the invoice value.

Yours faithfully,
for XYZ Co. Ltd.

ISSUE CONTROL

We now come to the final stage, namely, issues. Issues can be further divided into issues to consuming departments, and issues to outside suppliers for processing or conversion.

In both cases there are certain common system requirements. The first aspect is the control of issues. Issues are based on production programmes. Based on this and the bill of materials, work orders are printed, listing for each material, quantity to be issued against each component requiring that material. This automatically controls consumption because the work order gives details on quantity of materials to be issued and the corresponding quantity of components to be manufactured. So any materials requirement over and above that indicated in the work order quantity means excessive wastage and scrapping. Normally, stores personnel at junior levels are not authorised to issue beyond work order quantity. This automatically focusses top management's attention. Thus there is an inbuilt control. Sometimes materials are issued on loan basis. Proper control through stores registers must be ensured in such cases.

The second aspect is delegation of authority. Direct materials for which consumption norms can be established are controlled by work orders. For direct materials, such as fuel oil, electrodes, oxygen and tools, it is obvious that control should be based on past experience and suitable delegation. The stores assistant may authorise issues up to Rs 1,000. The Stores Officer's limit may be Rs 5,000 and so on. This establishes responsibility in controlling consumption. Ad hoc material requisitions are sometimes made. Periodically consolidated statements of such items must be prepared. Serial number controls are to be maintained and issues, as also receipts, must be posted in kardex so that stock balances are uptodate. The need for updating the records cannot be overemphasised as the stores record is the starting point of inventory management in all organisations. Imprest issues and replacement issues were mentioned with regard to tools stores in the previous chapter. When issues are made to outside suppliers controls have to be more formal and adequate enough to take care of payments and claims. Subcontracting often involves supply of raw materials by the subcontracting organisation and here stricter controls through despatch notes, consignment notes and purchase orders are exercised. In the Exhibits 22.2 and 22.3 are shown some of the stores documents discussed so far.

Good stores systems can greatly assist the Stores Manager in accurate stock status reports, timely detection of discrepancies, prompt clearance of Goods Inward notes to expedite bill payment, reduction in demurrages and losses in claims. For this propose a stores manual incorporating all the features has been prepared in many organisations. It is also essential that from time to time O&M studies should be carried out in the stores so that systems and procedures be streamlined.

23

Incoming Materials Control

The materials manager is very much concerned with the quality of incoming materials, as he is responsible for the supply of the right quality materials to the user departments, so that the quality of the end-products is ensured. For prudent materials planning, better buyer-seller relations and vendor rating, he should be aware of the inspection schemes and rejections, whether the items are inspected by the user departments or by an independent quality control section. In all organisations, the stores department inspects a large number of standard items, as well as detergents and soaps, and in a few organisations, the materials department also inspects raw materials and spare parts.

ACCEPTANCE OF STORES

All kinds of inspection, even counting the items, costs money. The use of statistical sampling schemes allows the management to decide upon tolerable levels for the risks of undesirably rejecting good materials and accepting bad materials when information is based upon sample inspection.

Quality control inspection is undertaken in industry in order to ensure that the products conform to the specified or pre-established standards. Such quality control inspection is carried out in the purchase of new materials, in the inter-departmental flow of materials or components from one department to another and in the final acceptance of finished products for distribution. Inspection of this kind helps to prevent sub-standard items from reaching the next user department and the final inspection prevents defective items from reaching the consumer. Through this process, it becomes possible to detect the various sources or factors which cause defective production and then to take appropriate remedial measures.

INSPECTION METHODS

Before a lot, consignment, batch, or shipment of items produced under homogeneous manufacturing conditions reaches the buyer, say, a manufacturing company buying raw materials or industrial components or a bulk buyer like the DGS&D, the following three alternatives are available to the buyer in making the purchase:

1. To accept the lot without any inspection;—this would depend upon the confidence placed in the supplier or on the assumption that the items have been inspected and passed at the supplier's end;
2. To inspect each and every one of the items in the lot; such cent per cent inspection is not always practicable, because of the considerations of time, cost, destructibility and fatigue;
3. To choose appropriate sample of the lot on the basis of sampling techniques and undertake planned inspection to identify items which conform to the standards and segregate items which are found to be defective; and, on this basis, to decide the acceptance or rejection of the lot; in this event the entire rejected lots may be returned to the supplier or it may be subjected to cent per cent inspection after which only the defective items may be returned to the supplier for replacement.

This last alternative method is called Acceptance Sampling. It is to be emphasised that the purpose of acceptance sampling is to determine the acceptance or rejection of the lot as a whole. The object is also to cumulate information about the manufacturing process, which will be fed back to the supplier in order to achieve and maintain a satisfactory standard.

AQL, LTPD AND RISKS

In the case of attribute characteristic, usually obtained by visual inspection or go-no-go gauge, the quality of the lot is expressed as percentage defective or average number of defects per item, whereas in the case of measured characteristic the quality of the lot is expressed in terms of the mean and the standard deviation of the measured characteristic. A sampling plan gives the procedure for acceptance and rejection of the lot with quantitative estimates of the risks to be taken by the producer and the consumer. The usual practice is for the buyer and the seller to determine a good or acceptable and a bad or rejectable lot in terms of acceptance quality level (AQL) and lot tolerance per cent defective (LTPD) defined as under: The consumer is willing to accept the lots whose quality is superior to a level of quality called the acceptance quality level, or AQL. Similarly, the consumer would like to reject lots inferior to a certain level known as the lot tolerance per cent defective or LTPD. These values are usually expressed in terms of per cent defective as 1/4 per cent or 1/2 per cent for AQL and 5 per cent or 10 per cent for LTPD.

As a result of sampling, it is likely that some 'good' lots are rejected and some 'bad' lots are accepted. The risk that a lot of items of acceptable quality will be rejected by the sampling as a result of a 'pessimistic looking sample' is known as the producer's risk. Similarly, the consumer's risk is the risk that a lot of items will be accepted by the sampling scheme as a result of an 'optimistic looking sample'. The consumer requires to be protected against rejection of an unreasonable number of good lots. Therefore, appropriate precautionary measures need to be taken. Some of the precautionary measures are described in the following paras.

RECTIFICATION—AOQL

If a lot, rejected by the sampling scheme, is merely sent back to the producer, the sampling scheme has little effect on the improvement of the material of the product. Further, rejection of the entire lot is not practicable, when sampling is in operation between different departments within an undertaking like foundry to machine shop, spinning to weaving, bottle washing to bottle filling, etc. The usual practice in such a situation is to screen the rejected lots item by item and get the defective items replaced by good ones. This is possible when inspection of the entire lot is economical and the test involved is not destructive. In such a case, the sampling plan gives a definite assurance regarding the quality of products accepted over a large number of lots. The diagrammatical representation of the rectification scheme is given in the Exhibit 23.1.

EXHIBIT 23.1
OPERATION OF A RECTIFICATION SCHEME

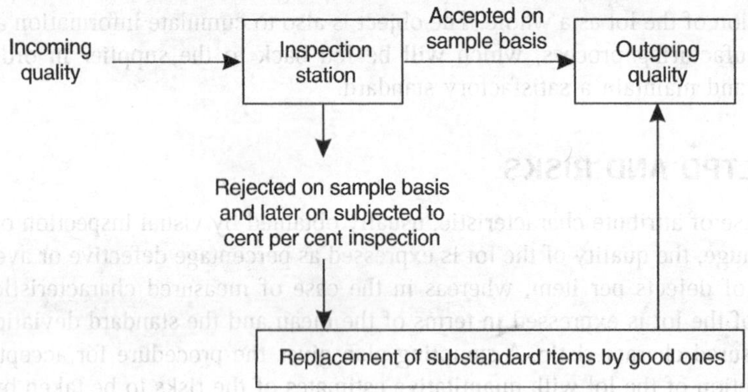

The average percentage or number of defective items in the accepted lots, after carrying out the inspection in the above manner, is known as the average outgoing quality (AOQ). A typical AOQ curve for sample size 100 and 'acceptance number' 2 is shown in Exhibit 23.2. The process is as follows:

100 items in the lot are inspected and if the number of defective items is found to be equal to or less than 2, the lot is accepted, otherwise the lot is rejected and is subjected to 100% inspection so that the substandard items are replaced. It can be noted from the chart that whenever the incoming quality is 'too good', the average outgoing quality is also 'too good'. On the other hand, if the incoming quality is 'too bad', a large number of lots are rejected on basis of sampling, inspected 100% and substandard items replaced by good items after screening. Hence the average outgoing quality is again 'too good'.

In between these two extreme situations, a limiting value or peak value of the average outgoing quality is obtained by the sampling scheme which is inferior in quality to the average outgoing quality of other lots. This maximum value of the average outgoing quality value is called the average outgoing quality limit (AOQL). This is shown in Exhibit 23.2.

EXHIBIT 23.2

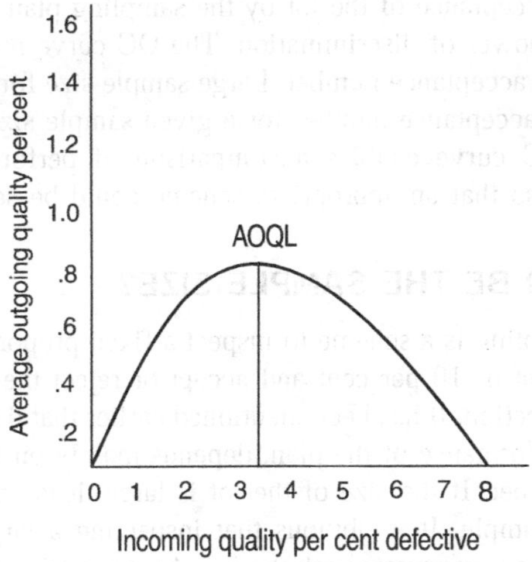

OPERATING CHARACTERISTIC CURVE (OC CURVE)

The protection afforded both to the consumer and the producer under any plan can be judged by its performance over a range of possible quality levels of submitted lots. The operating characteristic curve (OC curve) of a particular plan gives a picture of its operation. A typical OC curve for sample size 100 and acceptance number 2 is given in Exhibit 23.3. According to this scheme, 100 items are to be inspected and if the number of defects is 2 or less, the lot is accepted. Otherwise the lot is rejected (or rectified). On the horizontal axis, the lot quality per cent defective is plotted and on the vertical axis the proportion of

EXHIBIT 23.3

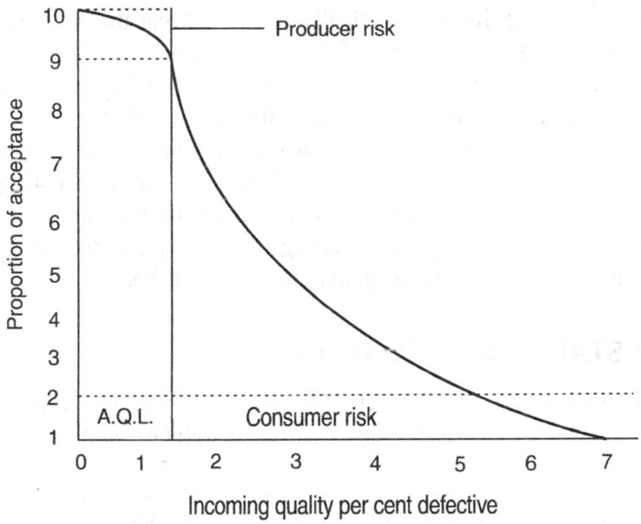

acceptance of the lot is given. For any given lot quality, the OC curve describes the proportion of acceptance of the lot by the sampling plan. The flatter the OC curve, the less its power of discrimination. The OC curve mainly depends upon the sample size and acceptance number. Large sample size for a given acceptance number or smaller acceptance number for a given sample size leads to a steeper OC curve. The OC curve enables a comparison of performance of different sampling schemes so that an appropriate scheme could be selected.

WHAT SHOULD BE THE SAMPLE SIZE?

Fixed per cent sampling is a scheme to inspect a fixed proportion of the items in a lot, say, 5 per cent or 10 per cent and accept or reject the lot on the basis of results of such inspection. It has been mentioned earlier that the O.C. curve giving a picture of the performance of the plan, depends mainly on the sample size and the acceptance number. If the size of the lot is large, it is not necessary to take a relatively large sample. It is obvious that inspecting a large sample involves effort which is not commensurate with the gain in terms of protection to the buyer or consumer. Thus it can be seen that fixed per cent sampling is not very useful whenever the lot size is larger.

For these small lot sizes (of about size 50 or so), it is not uneconomical to inspect all the items rather than depend upon the results of inadequate inspection based on fixed per cent sampling. This principle could be explained with the help of the following illustration: For instance, in a game of cards, it is a common experience statistically on about 38 per cent occasions, that a hand of six cards is free from any picture card—i.e., K, Q and J. From such small samples, involving 10 per cent sampling schemes of drawing 6 cards, out of 52 cards, one may erroneously infer that the lot is free from defects or any picture cards. Resorting to fixed per cent sampling is thus futile when the lot size is large or small. In actual practice the sample size for attribute characteristic is fixed between 100 to 500 through an agreement between the buyer and the seller based on concepts like AQL, AOQL, LTPD, etc. It can statistically be proved that the increase in gain is proportional only to the square root of the sample size.

Statistical methods are available to assess the reliability of tests based on sampling schemes and are useful in fixing the sample size depending on the desired accuracy of the estimate. Based on one or more of the concepts, AQL, LTPD, AOQL, consumer risk and producer risk, several tables indicating sampling schemes have been published. The more important and useful ones are (a) Military Standard 105D plans, (b) Dodge and Romig plans, and (c) Bowker and Goode plans. The salient features of these plans are presented below:

MILITARY STANDARD 105D PLANS*

These sampling procedures for inspection by attributes were developed by the US Department of Defence during World War II for its defence procurement and they

**Military Std 105D: Sampling procedures and tables on inspection by attributes,* US Department of Defence, US Government Printing Office, Washington DC, 1963.

have been subsequently improved. The focal point of these plans is the AQL value. For a given product characteristic, the supplier will be informed as to what the military agency considers to be an acceptable quality level. The defects are classified into three categories as critical, major and minor. Different AQL values are chosen for each category. Tables are available for AQL values ranging from .015 per cent to 1000, the former to indicate the per cent defective of lot quality and the latter to include defects per 1000 items. There are three levels of inspection and the lot sizes are grouped into 17 classes to include all possible lot sizes. Methods for altering the sampling inspection scheme, to become more strict in case of poor performance by the supplier, or when the 'process average' is worse than the specified AQL, and to reduce the amount of inspection when the process average is better than the AQL, are in-built in this scheme.

DODGE AND ROMIG PLANS*

These plans apply strictly to schemes that submit the rejected lots to 100 per cent inspection. A knowledge about the average fraction defective of the incoming material or estimated average per cent defective in the accepted lots during the recent past, defined as the process average, must be had before for using these tables. They are specially suitable for a manufacturer's internal inspection of his own products of one department, consumed in another. The consumer risk (or user department's risk) is fixed at 10 per cent for these plans.

The plans are based on AOQL and minimise the average amount of inspection for a given AOQL value and specified process average. Plans are available for different values of AOQL ranging from 0.1 per cent to 10 per cent. The corresponding LTPD values are also indicated. The plans based on LTPD similarly minimise the amount of inspection for a given LTPD value and a specified process average. Plans are available for LTPD values ranging from 0.5 per cent to 10 per cent. The choice between the two criteria, LTPD or AOQL depends on whether the emphasis is on segregating good lots from bad lots or on controlling the average quality reaching the assembly line.

BOWKER AND GOODE PLANS

Whenever the quality characteristic is measurable, it is possible to use the additional knowledge with advantage and suitable sampling schemes are available for this purpose. It is noted that in the variable inspection, the items are measured for comparing as to whether they meet the specification or not. In the case of attribute type, the items are usually classified as good or bad. Hence, for any designed degree of precision or discrimination between good and bad lots, compared to the attribute type of inspection, considerably smaller sampling plans are adequate as the variable inspection yields more information about the quality of the lot than the attribute inspection. The increased information about the quality of the lot may be used by the producer to improve the manufacturing process.

*Dodge and Romig, *Sampling Inspection Tables,* John Wiley & Sons, 1959.

Bowker and Goode have published a set of tables for variable inspection with the respective OC curves.* These plans are applicable when the quality of the item is measurable. Lot quality is judged in terms of per cent defective and the acceptance of the product is lot by lot. The item quality has either a single upper specification value or a lower specification or both. The two cases, when standard deviation is known and not known, are distinguished. The plans are specified in terms of two numbers *n* and *k*, *n* corresponding to the sample size and *k* a factor to be used in judging the acceptability of the lots.

CHOICE OF A SAMPLING PLAN

In each of the above cases, the concept of 'acceptance-rejection' or 'acceptance-rectification' on the basis of a single sample can be extended to multiple sampling. The decision-making process of rejecting the lot (or rectification of the lot) is postponed to one more stage in the case of double sampling and to further stages in the case of multiple sampling.

There are many features of a sampling plan which could be of interest, but the following are those which have been generally found useful for choosing a plan:

Type of quality characteristic whether measured or attribute type; the sample size; the acceptance number; the quality of batch which is almost certain of rejection if offered; the quality of batch which is almost certain of acceptance if offered; the quality of batch which has an even chance of acceptance or rejection if offered; the average outgoing quality limit; the batch size; the cost of inspection; the cost of rejecting an article; the cost of accepting a defective article; the frequency distribution of quality—that is to say, a knowledge of how often batches of each possible quality will be submitted.

Example. The materials manager of Madras Pharmaceuticals Limited is interested in introducing scientific sampling procedures to control visual defects like cracks, chips, breakages, etc., in the incoming bottles used for filling gripe mixture liquid. Initially his interest is confined to single sampling acceptance-rejection schemes. He was examining the following alternative schemes:

a. Inspect 10: accept if all are free from defects,

b. Inspect 50: accept if no defect or one defect is noted in the sample, and

c. Inspect 100: accept if 0, 1, 2 defects are found in the sample.

Which scheme should he adopt for procuring the bottles from the suppliers?

Solution. From published Poisson tables (Exhibit 23.4) for each of the three schemes (a), (b), (c) corresponding to typical lots containing 5 per cent, 10 per cent, 15 per cent defective could be obtained as follows:

Thus 'Scheme c' of inspecting 100 and allowing 2 defects in the sample minimises the acceptance of bad lots, and hence is preferable.

*Bowker and Goode, *Sampling Inspection Tables by Variables,* McGraw-Hill, 1952.

EXHIBIT 23.4
POISSON DISTRIBUTION

Number (of defectives) found in the sample	Approximate proportion (probability of occurrences in repeated sampling) Average value (of number defectives) expected in the sample									
	1.0	2.0	3	4	5	6	7	8	9	10
0	.368	.135	.050	.018	.007	.002	.002	—	—	—
1	.368	.271	.149	.073	.034	.015	.006	.003	.001	—
2	.184	.271	.224	.147	.084	.045	.022	.011	.005	.003
4	.061	.181	.224	.195	.140	.089	.052	.028	.015	.007
5	.015	.090	.168	.196	.176	.134	.091	.057	.034	.019
6	.001	.012	.050	.146	.161	.161	.149	.122	.091	.063
7	—	.003	.022	.060	.105	.138	.149	.140	.117	.090
8	—	.001	.008	.030	.065	.103	.130	.140	.132	.112
9	—	—	.003	.013	.036	.069	.101	.124	.132	.125
10	—	—	.001	.005	.018	.041	.071	.099	.119	.125
11	—	—	—	.002	.008	.023	.045	.072	.097	.114
12	—	—	—	.001	.003	.011	.027	.048	.073	.095
13	—	—	—	—	.001	.005	.014	.030	.050	.073
14	—	—	—	—	.001	.002	.007	.017	.032	.052
15	—	—	—	—	—	.001	.004	.009	.019	.035
16	—	—	—	—	—	—	.002	.005	.011	.022
17	—	—	—	—	—	—	—	.001	.003	.007
18	—	—	—	—	—	—	—	—	.001	.002
19	—	—	—	—	—	—	—	—	.001	.002
20	—	—	—	—	—	—	—	—	—	—
21	—	—	—	—	—	—	—	—	—	—

This table is useful to determine the control limits of attribute control charts and also to evaluate the performance of different acceptance sampling schemes.

A more detailed treatment of the design of single and multiple sampling plans is beyond the scope of this text. However, the interested reader may refer to standard quality control books such as Juran's *Quality Control Handbook*, 3rd Ed., McGraw-Hill, 1974.

EXHIBIT 23.5
PERCENTAGE OF ACCEPTANCE OF BAD LOTS

Scheme		*Lot per cent*	*Defective*	
Inspect	*Accept*	5	10	15
10	no defect	61	37	22
50	one defect	28	4	1
100	two defects	12	3	—

24

Stores Accounting and Stock Verification

STORES ACCOUNTING

In relation to the estimation of the cost of the product for pricing decisions, stores accounting assumes a key role. Material costing is very important in terms of the valuation of the cost of materials consumed by the production department as well as in terms of the estimation of the value of materials held in stock. We will discuss the materials costing under classifications of the receipt of materials, issue of materials, and of the stocks held at the end of the accounting period.

We will see the various methods used in costing and their limitations.

COSTING OF THE RECEIPT OF MATERIALS

The factors that are to be included in the building up of the cost of the materials received are material price, freight charges, insurance and taxes. Price usually refers to the price quoted and accepted in the purchase orders.

Prices may often be stated in various ways, such as net prices, prices with discount terms, free on board, cost insurance and freight, etc. For costing purposes we have to work out the actual cost incurred by taking price quoted by supplier as the basis, subtracting the discounts and adding any other expenses not covered.

The freight costs incurred in transporting the goods are usually collated under a separate head. Sometimes prices may include this element. Hence care should be taken to ensure that there is no double counting.

Goods in transit are mostly covered by insurance. All such insurance expenses must be calculated and added to the base cost and transportation cost.

Under the miscellaneous head, we need to classify costs incurred by way of customs duties, taxes, and packages. Such separate classifications give a better framework for cost control. In sum, we can say that cost of the materials received is equal to the price quoted less discounts, plus freight, insurance, duties, taxes and package charges. Very often such detailed classification helps in quicker analysis and effective control. Duty drawback statements, for example, are prepared by many organisations which want to avail of the exemption of duties in respect of the value exported. Such statements require a detailed break-up of various elements of cost. In the absence of the detailed classifications discussed above, it will become very difficult to prepare such statements.

COSTING OF THE ISSUES TO PRODUCTION

First in first out (FIFO), last in first out (LIFO), average cost, standard cost, base stock method, market price at the time of issue, latest purchase price, replacement or current cost are some of the methods used in costing the issues to production.

There are several other methods of costing also. We will discuss in following paragraphs some of the important and frequently used methods in detail.

FIFO

The assumption made here is that the oldest stock is depleted first. Therefore at the time of issue, the rate pertaining to that will be applied. This is logical in the case of items which deteriorate with time. Since actual prices are used, there cannot be any 'profit' or 'loss' in the pricing arrangements. In FIFO process, the value of the stocks held on hand is the money that has been paid for that amount, of stock at latest price levels and hence can straightaway be used in balance sheet, truly reflecting the value.

The limitations of FIFO process are that the process becomes unwieldy when too many changes in price levels are encountered and the fact that this method does not provide a satisfactory answer to costing returns from stores. A typical FIFO method of costing is explained below:

EXHIBIT 24.1

Date	*Receipts*			*Issues*			*Stock on hand*		
	Qty.	*Rate*	*Value*	*Qty.*	*Rate*	*Value*	*Qty.*	*Rate*	*Value*
1/1/75	—	—	—	—	—	—	—	—	—
1/2/75	1000	1.00	1000	—	—	—	1000	1.00	1000
14/2/75	2000	1.10	2200	—	—	—	1000	1.00	3200
							2000	1.10	
20/2/75	—	—	—	500	1.00	500	500	1.00	2700
							2000	1.10	
28/2/75	—	—	—	1500	1.00	500			
					1.10	1100	1000	1.10	1100

LIFO

The basic assumption here is that the most recent receipts are issued first. We can apply LIFO system to the example discussed earlier. The results could be as follows:

EXHIBIT 24.2

Date	Receipts			Issues			Stock on hand		
	Qty.	Rate	Value	Qty.	Rate	Value	Qty.	Rate	Value
1/1/75	—	—	—	—	—	—	—	—	—
1/2/75	1000	1.00	1000	—	—	—	1000	1.00	1000
14/2/75	2000	1.10	2200	—	—	—	1000 2000	1.00 1.10	3200
20/2/75	—	—	—	500	1.10	550	1000 1500	1.00 1.10	2650
28/2/75	—	—	—	1500	1.10	1650	1000	1.00	1000

Under LIFO system. In a period of rising prices, latest prices are charged to the issues, thereby leading to lower reported profits and hence savings in taxes. When there are wide fluctuations in price levels, LIFO tends to minimise unrealised gains or losses in inventory. However, LIFO systems have the same disadvantages as that of FIFO systems.

Average cost

In this method, the issues to the production department are split into equal batches from each shipment at stock. It is a realistic method reflecting the price levels and stabilising the cost figures. The average cost method when applied to our example will produce the following results:

EXHIBIT 24.3

Date	Receipts			Issues			Stock on hand		
	Qty.	Rate	Value	Qty.	Rate	Value	Qty.	Rate	Value
1/1/75	—	—	—	—	—	—	—	—	—
1/2/75	1000	1.00	1000	—	—	—	1000	1.00	1000
14/2/75	2000	1.10	2200	—	—	—	3000	1.067	3200
20/2/75	—	—	—	500	1.067	533	2500	1.067	2667
28/2/75	—	—	—	1500	1.067	1600	1000	1.067	1067

Thus, here the rate is arrived at by dividing the total cost by the number of items. Then, this rate is applied to the issues to production. As more purchases are made, a new average is computed and this average is applied to the subsequent issues. Generally these three methods can be compared in the following two cases:

EXHIBIT 24.4

(i) Rising price levels:

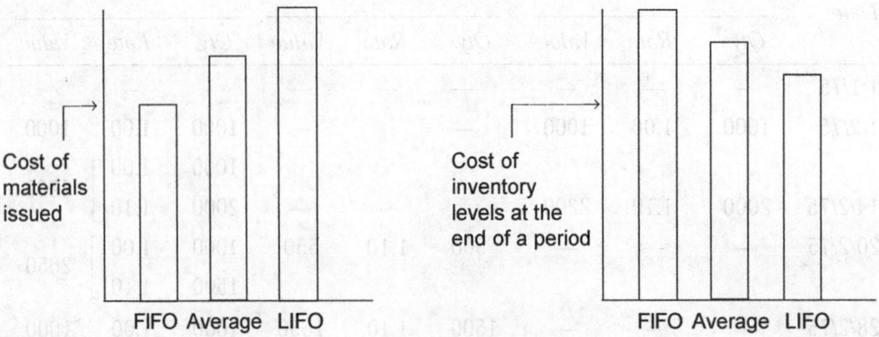

(i) Falling price levels:

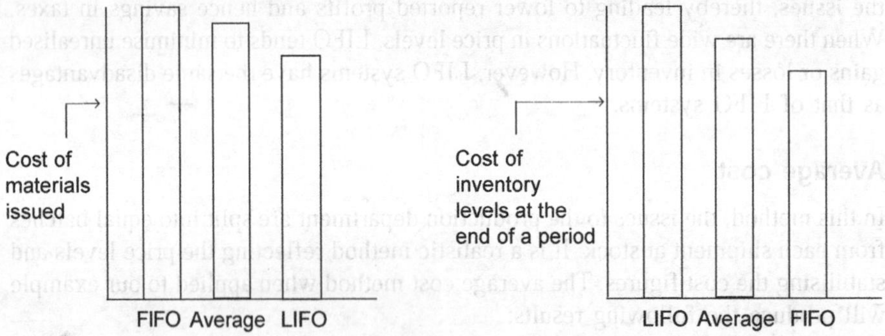

From the above exhibit, we can easily infer that depending upon the method used the profit and loss for any period can considerably vary especially when large value of materials are consumed.

Market Value

This method is also known as replacement rate costing. Here the materials that are issued are costed at the market rate prevailing at the time of issue. It, therefore, follows that when prices increase the stocks on hand is continuously under-estimated, because receipts are costed at actuals and issues at higher rates. Conversely, when the prices are falling, the stock on hand is grossly overestimated. This may in turn lead to writing off huge amounts to make it realistic. Besides, this system requires continuous monitoring of market rates for all materials and hence is very unwieldy.

Standard Costs

Here, a standard rate is determined based on detailed analysis of market prices and trends. This standard rate is kept fixed for a definite period of six months or more. During this period costing is done on the basis of this standard rate, irrespective of the actual rates. At the end of the period, a review is done and fresh standards are set for a further period of six months.

Efficient use of materials is truly reflected by adopting this method as the accounting is divorced from fluctuations in rates. Further it is not necessary to obtain fresh rates at every point of time. This means greater clercial efficiency and quicker estimation of costs. However, in this method also at the time of rising prices, the stock on hand is underestimated, and at the time of falling prices, the stock on hand is overestimated.

Costing the Closing Stock

Generally the guideline used here is that either the market price or stock at cost is to be used, whichever is less. The main factors which determine the cost of closing stock are price levels, obsolescence and deterioration.

When the prices fluctuate, the kind of system used for evaluating the cost of issues to production affects the costing of closing stock. As we saw earlier, each system of costing tends to undervalue the stocks during periods of decreasing price levels. So, a provision has to be made to account for such variations from the actual value of the stocks.

While evaluating the closing stock costs, some stock items, such as machinery spares, tools, etc., tend to become obsolete earlier than others. So, in evaluating the cost of stock, at the close, a provision must be made to account for such obsolescence. This is based on past experience and is usually worked out as a percentage of the total stock value.

Many stock items deteriorate with time due to limited shelf life or inadequate precautions while storing. When the bulk of the items belong to this category, then losses due to deterioration can be very high. Provision must be made for this factor also in evaluating the stock at the end of the period.

In rare cases, stock may appreciate with time. Examples are liquor in the process of maturing and timber in the process of seasoning.

STOCK VERIFICATION

It is the process of physically counting, measuring or weighing the entire range of items in the stores and recording the results in a systematic manner. The purposes served by stock verification are as follows:

To reconcile the stock records and documents for their accuracy and usefulness. To identify areas which require more disciplined document control. To back up the balance sheet stock figures; and to minimise pilferage and fraudulent practices.

Stock verification is usually carried out by the materials audit department, reporting to either the materials manager or the internal audit. One person is usually given the exclusive responsibility with adequate facilities and authority.

Physical verification can be carried out periodically or on continuous basis

PERIODIC VERIFICATION

Under this system, the entire cross-section is verified at the end of one period, which is usually the accounting period. In big organisations this is not achieved in a day and usually several days are taken to complete this task. As no transactions can take place during the verification, this could pose some problems. Physical verification requires careful planning and execution. The various steps involved are detailed below:

A detailed programme should be chalked out giving complete breakdown of the process storewise and itemwise. This should be done in consultation with the materials management and finance departments. Necessary stock verification cards and checksheets must be prepared in adequate amounts. All material audit personnel must have clear-cut instructions on their jobs and schedule for proper accountability. During the verification process all transactions must be stopped. In other words, there should not be any receipts or issues. All stock-verification cards should be serially numbered for easy reference and control. Separate provisions must be made available for items which are damaged or deteriorated. Selected areas and items must be allocated to each stock-taking person so that orderly completion of the job without duplication or omission is ensured. It will be necessary to separately verify items which are under inspection, items sent out to suppliers for processing and stocks at various stock yards.

EXHIBIT 24.5

Store location.. Serial No. ....
Part number.. Date. ....
Part description..
Unit of quantity NO/METRES/KGS/LITRES
Rate.. Quantity on hand...............
Value of stock on hand ....
Audited by S/d
Checked by S/d

Remarks

Damaged Deteriorated Others

Such sheets as shown in Exhibit 24.5 are prepared for each item and values are worked out for different classifications. The total of such values gives the value of the stock on hand as verified. Then this is tallied against the book figures or stock records. Discrepancies, if any, are noted down. Minor discrepancies are taken care of by correcting the stock records. Any major discrepancies need further analysis so that causes can be identified and remedied. Allowance regarding acceptable margins of tolerances for conversion, weighing and measuring, as well as for evaporation, must be clearly laid down. Top management's sanction can then be sought for writing off deficiencies or valuing surplus.

CONTINUOUS VERIFICATION

Under this system, verification is done throughout the year as per a pre-determined plan of action. A-items may be verified thrice a year, B-items twice a year and C-items once a year. It, therefore, presupposes that a perpetual inventory record for each item is maintained showing all transactions so that reconciliation can be done. The advantages here are:

Work can be independently carried out by materials audit department staff. Investigation with regard to discrepancies are spread over the year and hence detailed analysis is possible. Final accounts can be prepared expeditiously if continuous verification is done as per plan. There is no need to 'freeze' the entire operations of the stores as verification is done throughout the year based on perpetual inventory records. Any time stock records are more up-to-date when compared with the periodic verification system.

PROCESS OF VERIFICATION

Items are verified by counting in the case of bearings, by weight in the case of sheets, by measuring in the case of lubricants and so on. However, when large stocks of items such as sand, scrap and ore fuel need to be verified, it is based only on estimates as the question of exact measurement is ruled out. In the actual process of stock verification, the stores personnel should be involved, as they intimately know the locations of various items which results in quicker identification of items. For instance, some items may be located in many places. By virtue of their experience, only stores personnel will be able to locate them. So the material audit people will have to work in close coordination with them. Discrepancies must be discussed with Stores so that any omissions may be rectified and then only should they be reported to top management. Major discrepancies may require a re-verification. Such discrepancies may be due to pilferage on a large scale, wrong posting of records and loose documents control. They require careful analysis and immediate corrective measures.

After discrepancies have been noted, stock adjustments must be made using standard stock adjustment documents duly signed by the appropriate authority. A typical stock adjustment form is shown below:

EXHIBIT 24.6

DISCREPANCY VOUCHER

Date of verification:	Serial number ....
Part number:	Serial no./ref. no. ....
Part description:	Stock verification sheet
Location code:	....
Quantity as per record:	
Quantity on verification:	
Discrepancy: Amount: Value:	Surplus
	Deficient
Prepared by S/d	
Approved by S/d	

After the approval, the stock records can be corrected. Surprise checks and verifications are made by materials audit department to detect any fraudulent acts. Material audit plays the role of a watchdog of stores, pointing out weak areas and remedying them. It assists in accurate records-keeping and smooth finalisation of annual accounts.

25

Obsolete, Surplus and Scrap Management

DEFINITIONS

Obsolete items are those materials and equipment which are not damaged and which have economic worth but which are no longer useful for the company's operation owing to many reasons such as changes in product line, process, materials, and so on. Surplus items are those materials and equipment which have no immediate use but have accumulated due to faulty planning, forecasting and purchasing. However, they have a usage value in future. Scrap is defined as process wastage, such as turnings, borings, sprues and flashes. They may have an end-use within the plant. In any case they have a market value. We have already seen that as much as nearly Rs 2,500 crores are tied up in obsolete and surplus inventory in India.

From the above definitions we can list the reasons for the generation and accumulation of obsolete, surplus and scrap items:

1. *Changes in product design.* This may lead to some items getting invalid so far as the final product is concerned. Hence, the entire stock of such items becomes obsolete.

2. *Rationalisation.* Sometimes raw materials are rationalised so as to minimise variety and simplify procurement. The rationalisation process renders some items as surplus or obsolete.

3. *Cannibalisation.* When a machine breakdown occurs, sometimes it is rectified using parts of an identical machine which is not functioning due to various reasons. This process of "cannibalisation" is not uncommon in many project-based industries. When continued unchecked, this results in obsolete and scrap items.

4. *Faulty planning and forecasting.* The marketing department may have projected a sales forecast which might be on the higher side. Any materials planning has to be based on sales forecasts and this could result in surplus items. Wrong indenting by the user departments also leads to accumulation.

5. ***Faulty purchase practices.*** Sub-optimising decisions like buying in bulk to take care of discounts and transportation economy without taking into account factors such as shelf life, storage space requirements and technological changes once again lead to the accumulation of surplus and obsolete stocks.

6. ***Other causes.*** Many items are held as insurable spares for many years without any consumption. Faulty store-keeping methods, without adequate preservation, lead to spoilage. Inferior materials handling, improper codification and poor manufacturing methods also result in obsolete, surplus and scrap items. Poor maintenance of machine tools may result in excessive tool wear and greater process scrap.

IDENTIFICATION AND CONTROL

The combing process of combining the stock records and movement analysis has been found very effective in locating such stocks in the total inventory. Stock issue cards should be combed and items which have not been consumed (non-moving) for a period of one year must be isolated. A list of such items and their value in terms of money and time must be made. Similarly, such lists must be prepared for items which have not moved for 2 years, 3 years, 5 years and above. Such lists can then be put up to top management for disposal action. Care must be taken to prepare a separate list of imported spares and insurance items. Such combing and movement analysis must be done on a continuous basis. A typical movement analysis statement is shown in Exhibit 25.1.

EXHIBIT 25.1

Sl. No.	*Part number description*	*A B C*	*Last date of issue*	*Stock on hand in number of days consumption*	*Value of orders on hand*

(The statements will be prepared in such a way that items with no issue for a long time will appear on top of the statement. This will help in cancelling the orders and tapering off the stock on hand.)

Whenever changes in production programme, design and product lines are contemplated, a senior executive from materials management must definitely be kept in the picture. This helps in several ways. He is in a position to inform top management of the amount of stock of materials on hand that are likely to be rendered obsolete if and when the changes are introduced. This could even guide the management as to when the changes could be made so that the existing stock can be consumed in full. The materials manager in turn can freeze further orders for such materials and try to negotiate with the suppliers to take back the stock. For some items he can introduce the buy-back clause wherein the supplier takes back items not consumed within a specific period. For new items which may be required, he can try to develop sources and place orders so that changes can be expeditiously introduced. All this highlights that a close coordination is required in order to avoid stockpiling obsolete and surplus items. Selective control based on ABC analysis, accurate forecasting techniques and proper preservation minimises such accumulation. In the case of storage of perishable items* the Central Warehousing Corporation has devised a scheme to enlarge its scope to cover certain specialised lines of storage. The schemes drawn up by the corporation include setting up of a chain of cold storage plants for potatoes, and fruit, and storage facilities for certified seeds. The other specialised area which the Central Warehousing Corporation is likely to enter is the maintenance of buffer stocks of certified seeds to the extent of 50,000 tonnes in four states.

Many organisations have introduced formal documentation in introducing changes in design or product. It is called the "Effective Point Advice". This is popularly known as EPA. Here, the proposed changes, details of new materials and products required, details of materials and products which will be invalid/ obsolete when the change occurs and the approximate date when the change is expected to be introduced are detailed and circulated to concerned departments. EPA thereby helps in tapering off the stocks of "invalid" items, cancellation of orders for such items, placing orders for buying and/or manufacturing new items and related activities. EPA systems help in better coordination for profitable introudction of changes with minimum "side effects" such as the accumulation of obsolete items.

The reclamation of scrap has not attracted the attention of the top management in Indian industry. Optimal utilisation of scrap would allow conservation of the use of scarce natural resources, such as iron ore. This is because scrap is an important element which goes into the manufacture of steel and castings. In the following paragraphs we will see how scrap can be managed in industry.

It is possible to salvage scrap for usage within the firm in some cases. Press parts are normally made from sheets and plates. The off-cuts generated during such process can be profitably utilised in making smaller press components such as washers. Rationalising the supply size, changes in process such as reversing the dies may result in minimisation of scrap. Big organisations have a full-fledged scrap salvaging department. These departments segregate the scrap into categories, like turnings, borings, plate cuttings, end-pieces of billets, purchasing, etc. Colour coding the scrap is also done to avoid the mixing of different categories of scrap. It is also advantageous to the end-users in each category and supply of scrap at the appropriate time for production. Often informing the production department at

*_The Economic Times_, 13-9-1975.

the right time about the volume of scrap generated enables prevention of excessive scrap through timely changes in production methods, toolings and materials.

DISPOSAL OF SCRAP

Disposal of scrap when handled in an imaginative manner can result in handsome returns to the organisation. An effective disposal requires a compact disposal organisation reporting to the materials manager, continuous market survey on the prices of various categories of scrap generated in the plant and constant touch with the industries which generate similar scrap and with the end-users.

Disposal action follows when the scrap cannot be utilised within the organisation. In practice, it has been found that it is profitable to dispose the scrap directly to end-users rather than to middlemen who normally form a cartel of their own which leads to lower returns. Before disposal action it is essential that the scrap is segregated according to metal, size, etc. When the scrap is mixed, the return is even lower than the lowest element in the mixture. This is because the buyer of scrap will have to segregate it at an extra cost. A cursory analysis of scrap prices will reveal that sheet and plate cuttings will fetch around Rs 3,000 a tonne whereas turnings and borings only Rs 600 a tonne. Also when costly scrap such as copper, aluminium and tungsten are involved it is imperative that they are segregated as returns are huge and price levels are different. Since scrap is generated process-wise, it comes out in a segregated condition and there should be no difficulty in sorting.

Many methods are used for disposal. Two methods which are frequently used are by auction and by tender.

Parties in both the cases are normally required to inspect the scrap in the scrap yard and deposit earnest money. Very often the company insists on a basic price depending upon the category of scrap. The disposal section works, in this aspect, in close coordination with the finance department. In many cases the disposal section may try to enter into a long term contract with end-users such as steel plants.

Many companies have found to their displeasure scrapped components appearing in the market and competing with their parts as "original equipment". This is the price which organisations pay for not dismantling and disfiguring the scrap before disposal. Automobile spare parts and bearings especially are prone to such dangers. For this purpose some organisations go to the extent of requesting vehicle users to demolish filters and plugs before scrapping them. This is a very important aspect.

In view of the paucity of raw materials and shortage of credit, it is necessary that optimum usage of materials is made and funds tied up in obsolete surplus and scrap items minimised. This is only possible when top management shows commitment and support. The employees of the organisations are naturally the best people to suggest improvements in materials, processes and new end-users for scrap. It is they who can minimise the accumulation of scrap through coordination. Therefore, top management should work out formal reward systems to promote employee participation in this matter. A few organisations have suggestion box schemes which pay rich dividend to the organisation. Employees, too, get rewards and recognition in the process.

26

Value Analysis

ORIGIN

Value engineering or value analysis had its birth during the Second World War. Needless to say, this was a fertile period for its growth due to world-wide shortage of essential materials. Lawrence D. Miles was responsible for developing the technique and naming it. Through its application, G.E.C. saved nearly 70 million dollars and after that it was a question of time before the technique caught on and spread like wild fire. In recent years, it has also been known as "molecular engineering" or "vertical thinking".

DEFINITION AND SCOPE

Value analysis is defined as "an organised creative approach which has, as its objective, the efficient identification of unnecessary cost—cost which provides neither quality nor use nor life nor appearance nor customer features". Value analysis focuses engineering, manufacturing and purchasing attention to one objective—equivalent performance at a lower cost. In other words, functional performance remains unaffected. Other techniques are as follow:

1. *Organised cost reduction analysis* by which accountants analyse costs in products or procedure, the ranking of their elements in a descending order of magnitude and in informed challenge to each, starting with the most important.

2. *Development as practised in many firms* where it follows the realisation of a design and combining through the specifications to remove costs by substituting standards and relaxing tolerances and finishes.

3. *Purchasing analysis* which systematically searches for cheaper bought-out components or services, probably by seeking alternatives or a fresh basis for price negotiation.

4. *Method study.* The part of work study that studies and analyses work of any kind in a disciplined manner that leads to improvement in methods.

The area of application of value analysis is best given by the schematic diagrams—Exhibits 26.1 and 26.2. Value analysis is hence concerned with the costs added due to inefficient or unnecessary specifications and features. It makes its contribution in the last stage of product cycle, namely, the maturity stage. At this stage Research and Development no longer make positive contributions in terms of improving the efficiency of the functions of the product or adding new functions to it. The prime task of the manufacturer becomes one of tackling the value part of the job so well that product leadership will be maintained.

EXHIBIT 26.1

Total cost of product including		Minimum that must be spent on product
	Work added by unnecessary design and specification features.	Value analysis zone....
	Work added by Inefficient methods of manufacturing.	Method used study zone
	Overhead because men/ machines are idle	Production control zone

EXHIBIT 26.2
PRODUCT LIFE CYCLE

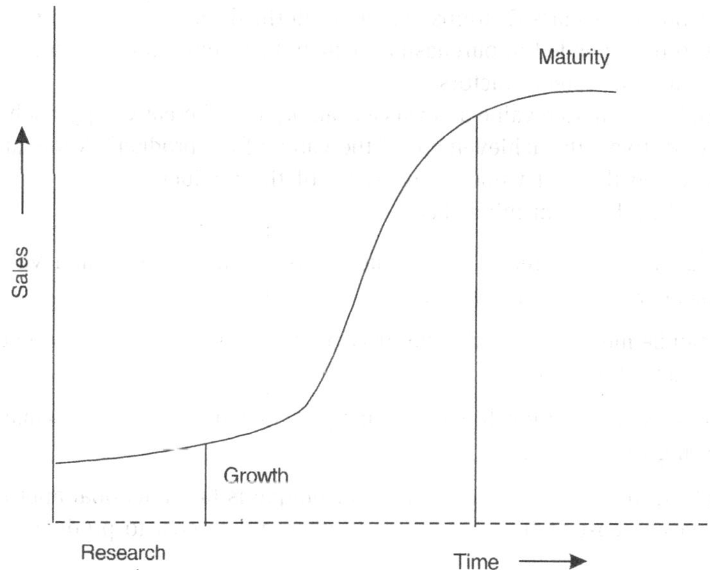

The other extra costs shown in Exhibit 26.1 that of overheads and work added by inefficient methods are tackled during the growth stage in Exhibit 26.2. As the sales of a product pick up, the problem of overhead is automatically eliminated. The work methods improve as a process of learning from repetitive

The next phase deals with analysis and evaluation. In this phase suggested alternatives are analysed and evaluated regarding costs, probability of applicability, etc. Hurdles to the successful acceptance of the alternatives generated are studied and methods or approaches to overcome them developed.

During the programme planning phase the programme or plan for studying the various alternatives in greater depth, etc., is laid down.

We then have the execution process. During this phase all the alternatives, with the costs and benefits, are studied according to the plan or programme, to finally arrive at the best alternative.

The final phase deals with the presentation and implementation phase. During this phase all relevant data are presented to the decision-making body or authority. The accepted ideas are then implemented and closely followed to study the actual benefits derived.

ORGANISATION FOR VALUE ANALYSIS

The definition of value analysis emphasised that it is an organised creative approach. It also indicates that value analysis focuses the attention of Design, Purchase, Packaging, Production, Finance (Costing) and Sales departments to one objective—equivalent performance at a lower cost. This means that representation from each of these departments is essential to form a value analysis team. The interests of the representatives from these departments are obviously conflicting. It is well known that an atmosphere of positive conflict induces a lot of creativity in the individual. The team members will be at their creative best.

The methodology of value analysis suggests that it is a systematic and long process. It is not possible for departmental representatives to devote so much of their time to value analysis alone. The value analysis team should therefore be headed by an executive specifically earmarked for this purpose. The other members could be associated on a part-time basis. This person should be close to top management and must be effective in the implementation of the team's findings. This will keep the team members motivated.

VALUE ANALYSIS BY MATRIX SYSTEMS

The Matrix System of value analysis is a slight variation of value analysis, in that the most effective material is chosen according to numerical coefficients. The matrix system is applied to the various alternatives which are generated by the value analysis technique to fulfil the required functions in order to choose the best among them. The matrix system is based on "emphasis coefficients" and "emphasis factors".

Emphasis coefficients indicate the weightages to be attached to the various functions of factors. These coefficients indicate how efficiently the alternatives (items) perform each function. This means that all the alternatives are weighed against each function or factor. The number of sets of emphasis factors is equal to the number of functions or factors. The sets of emphasis factors are then multiplied by the emphasis coefficients and the material with the highest value is chosen.

Indian Examples In India several public and private sector undertakings have applied value analysis techniques successfully for cost-reduction and import substitution purposes. Some are discussed below:

Substitution of solid aluminium extrusions for hollow rolled steel sections in wind shield frames in cars. This effected a saving of Rs 80,000 per annum. The use of aluminium in place of copper for manufacturing cables has helped conserve valuable foreign exchange.

The country is now self-sufficient in aluminium and the imports of copper have come down by about Rs 8 crores. Aluminising in place of galvanising has helped reduce the imports of zinc.

The use of duplex board cartons in place of tin containers for milk powder has reduced the cost of packaging.

27

Material Handling

Material handling in Indian industries accounts for nearly 40 per cent of the cost of production. It is obvious that no value is added to the end-product through material handling. However, poor material handling may result in delays leading to idling of equipment. Nearly 50 per cent of the production cycle time in many industries is spent on handling materials. Materials handling can be defined as the function dealing with the preparation, placing and positioning of materials to facilitate their movement or storage. Thus the function includes every consideration of the product except the actual processing operation. In many cases, the handling is also included as an integral part of the process. Through scientific material handling considerable reduction in the cost as well as in the production cycle time can be achieved.

INFLUENCING FACTORS AND CONTROL

Materials handling cannot be viewed in isolation. Many factors like plant layout, processes, nature of raw materials and products influence the material handling system. Therefore, an integrated approach will have to be taken in the case of material handling.

It must be appreciated that material handling operations encompass suppliers, stores, inspection, manufacturing, packaging, warehousing and transport to consumer. The objective, therefore, must be to obtain maximum overall effectiveness in material handling. Many decisions have a significant effect on the effectiveness of the material handling system. We will discuss some of the important decisions.

Layout decisions are normally taken on technological considerations. Very rarely are such decisions taken in consultation with materials management and materials handling departments. Such a consultative process could result in savings in terms of reduction in handling costs, reduction in investment on handling equipments and a reduction in the production cycle time. In such a consultative process, movement of materials could be studied right at the outset from the receipt of raw materials stage to that of despatch of finished goods to warehouses. Many organisations have used industrial engineering techniques such as flow process charts and string diagrams to design effective integrated material handling systems. When layout and handling are designed in an integrated manner, valuable insights can be developed on the selection of handling equipments, number of equipments and maximum utilisation of floor space. In the two basic layouts, viz., product layout and process layout, it has been found that material handling is kept at a minimum in the case of product layout. This is because product layout is designed as per the sequence of operations leading to smooth flow of materials. Also, machines are so located as to minimise the handling between successive processes. However, process layout results in lower investment in capital equipment and better utilisation. Many compromise layouts can be worked out to suit specific requirements. A good layout should be such that there is no congestion of materials in process inventory. Transportation lines and bottlenecks must be minimal.

EQUIPMENT

Broadly material handling equipments can be classified into two categories, namely: (a) Fixed path equipments, and (b) Variable path equipments. It is gratifying to note that India is manufacturing many of these equipments. For instance, the largest conveyor system in Asia is to be installed at Neyveli Lignite Corporation by Elecon Engineering Company Limited. This is a conveyor system of 1,800 and 2,000 mm belt widths and 15 km length with three crawler mounted trippers weighing 500 tonnes each, to handle over 11,000 tonnes burden per hour in open cast mines, and worth Rs 240 million.* The earth moving equipment industry manufacturing shovels, excavators, crawler tractors, dumpers, motorised scrapers, loaders, road rollers, etc., has also made substantial progress. India had exported materials handling equipment worth 77 crores during 1974–75.

Fixed path equipments are those equipments which move in a fixed path. Conveyors, monorail devices, chutes, and pulley drive equipments belong to this category. A slight variation in this category is provided by the overhead crane, which though restricted, can move materials in any manner within a restricted area by virtue of its design. Over-head cranes have a very good range in terms of hauling tonnage and are used for handling bulky raw materials, stacking and at times palletising.

Variable path equipments have no restrictions on the direction of movement although their size is a factor to be given due consideration. Trucks, forklifts mobile cranes and industrial tractors belong to this category. Forklifts are available in many ranges, they are manoeuvrable and various attachments are provided to increase their versatility.

Having discussed the variety of equipments available we will not see the factors which have to be evaluated in the selection of material handling equipments. The cost factors include investment cost, labour cost, anticipated service hours per year, utilisation, unit load carrying ability, loading and unloading characteristics, operating costs and the size requirements. Other factors to be considered are source of power, conditions where the equipment has to operate

**Economic Times*, September 5, 1975.

and such other technical aspects. Therefore, choice of an equipment has to be made in the light of the factors discussed above. Even in an on-going organisation it is possible to improve the material handling system through work study techniques. They usually result in improving the ratio of operating time to loading time through palletising, avoiding duplicative movements, etc. Obsolete handling systems can be replaced with more efficient equipments.

GUIDELINES

The following guidelines are invaluable in the design and cost reduction of the material handling system:

1. As material handling adds no value but increases the production cycle time, eliminate handling wherever possible. Ideally there should not be any handling at all!
2. Sequence the operations in a logical manner so that handling is unidirectional and smooth.
3. Use gravity wherever possible as it results in conservation of power and fuel.
4. Standardise the handling equipments to the extent possible as it means interchangeable usage, better utilisation of handling equipments, and lesser spares holding.
5. Install a regular preventive maintenance programme for material handling equipments so that downtime is minimum.
6. In selection of handling equipments, criteria of versatility and adaptability must be the governing factor. This will ensure that investments in special purpose handling equipments are kept at a minimum.
7. Weight of unit load must be maximum so that each "handling trip" is productive.
8. Work study aspects, such as elimination of unnecessary movements and combination of processes should be considered while installing a material handling system.
9. Non-productive operations in handling, such as slinging, loading, etc., should be kept at a minimum through appropriate design of handling equipment. Magnetic cranes for scrap movement and loading in furnaces, combination of excavators and tippers for ore loading and unloading in mines are examples in this respect.
10. Location of stores should be as close as possible to the plant which uses the materials. This avoids handling and minimises investment in material handling system.
11. Application of OR techniques such as queueing can be very effective in optimal utilisation of material handling equipments.

12. A very important aspect in the design of a material handling system is the safety aspect. The system designed should be simple and safe to operate.
13. Avoid any wasteful movements—method study can be conducted for this purpose.
14. Ensure proper coordination through judicious selection of equipments and training of workmen.

The material handling institute in Pittsburgh has brought out interesting pamphlets with regard to 'do's and don'ts' of materials handling, and they contain very interesting points.

EVALUATION OF MATERIAL HANDLING

The effectiveness of material handling system can be measured in terms of the ratio of the time spent in the handling to the total time spent in production. This will cover the time element. The cost effectiveness can be measured by the expenses incurred per unit weight handled. It can be safely said that very few organisations try to collate the expenses and time in this manner so as to objectively view the performance and to take remedial measures. Some of the other indices which can be used for evaluating the performance of handling systems are listed below:

Equipment utilisation ratio is an important indicator for judging the materials handling system. This ratio can be computed and compared with similar firms, or in the same firm over a period of time.

In order to know the total effort needed for moving materials, it may be necessary to compute Materials Handling Labour ratio (MHL). This ratio is defined as under:

$$\text{MHL} = \frac{\text{Personnel assigned to materials handling}}{\text{Total operating work force}}$$

In order to ascertain whether the handling system delivers materials to work centres with maximum efficiency, it is desirable to compute Direct Labour Handling Loss ratio. The ratio is:

$$\text{DLHL} = \frac{\text{Materials handling time lost by direct labour}}{\text{Total direct labour time}}$$

The movement's operations ratio which is calculated after dividing total number of moves by total number of productive operations, indicates whether the workers are going through too many motions because of poor routing. It should however be emphasised that the efficiency of materials handling mainly depends on the following factors: (i) efficiency of handling methods employed for handling a unit weight through a unit distance, (ii) efficiency of the layout which determines the distance through which the materials have to be handled, (iii) utilisation of the handling facilities, and (iv) efficiency of the speed of handling.

In conclusion, it can be said that an effective material handling system depends upon tailoring the layout and equipments to suit specific requirements. When a large volume has to be moved from a limited number of sources to a limited number of destinations then fixed path equipments like rollers, belt conveyors, overhead conveyors and gauntry cranes are preferred. For increased flexibility varied path equipments are preferred. In practice, as we have seen in this chapter, specific requirements will have to be the guiding factor.

A Forklift truck which can be converted as a stoker.

28

Transportation and Traffic Management

IMPORTANCE

Transportation of materials from sources located in different parts of the country to the user's plant is of critical importance in the field of materials management. It can be safely said that the cost of transportation of materials account for nearly 10 to 20 per cent of the cost of the materials purchased. In fact, this will be much higher in the case of industries where suppliers are located in remote areas and the industry depends on heavy import. In some cases such as ores, the transportation charges are greater than the cost of the ores. Unfortunately this area has not been given due recognition and very often delegated to lower levels in the organisation. Very few organisations have realised its importance in terms of cost and customer service and integrated this function under materials management. Quite often, it will be noticed that the mode of transportation of materials is left to the suppliers. The Materials Manager should constantly seek improvements in transportation with a view to cutting down transportation costs, achieving greater speed and better quality of service. Materials movements may be local, regional and often national or international. As the organisations grow in size, transportation becomes very crucial and effectiveness depends upon the knowledge of alternatives, their rates, schedules, merits and demerits. Careful planning and analysis with appropriate controls often yield major economies in the operation, resulting in better service and higher profitability.

The important aspects in traffic and transportation management are choice of mode of transport, route selection, rate verification and auditing, management of claims and lost shipments as well as application of linear programming to minimise transportation costs.

Let us now consider each aspect in detail. While dealing with the choice of mode of transport the avenues that are open to the materials managers are shipping, rail, road and others.

TRANSPORT BY SHIPPING

This is generally resorted to in case of imports. The terms used in shipping for imported items are given in Exhibit 28.1. At times this method is used for movement of materials from one part to another within the country as well. For example, coal is moved from Calcutta port to Madras by ships. By and large transportation by sea is economical when distances involved are very large. Nowadays bulk tankers are built which have a DWT of over 10,000 and more. They are used for carrying such vital commodities as foodgrains, oil, etc. The increased use of shipping makes us look into the need for more berthing facilities at our ports. Ports at Bombay, Calcutta, and Madras need more berths so as to effectively handle the increasing number of ships which call at them every day. The material handling facilities at ports also require improvement. Bulk ore handling equipments and other sophisticated material handling devices are required for rapid unloading. On a number of occasions it has been noticed that the unloading process is painfully slow due to inferior handling devices and labour unrest at ports. This, coupled with the absence of an integrated inland transport system connecting the port with the plants and warehouses in the hinterland, may result in industries starving for materials when they are at port and foodgrains not reaching scarcity areas. It may also be pointed out that customs regulations and formalities if streamlined could result in reduction in administrative lead time and expeditious clearance of materials. Storage facilities at ports are necessarily hit by the factors mentioned above. The facilities can be better utilised by focussing attention on the areas discussed.

EXHIBIT 28.1

TERMS USED IN IMPORTS

1. *F.O.B.* (Free on board)

   Seller's responsibility:
   (a) Place goods in carriers;
   (b) Obtain bill of lading from carriers;
   (c) Answerable for damages and losses till the goods are placed in carriers and bill of lading (clean) has been furnished by the carrier.

   Buyer's Responsibility:
   (a) Movement of goods once they are on board:
   (b) Payment of transportation charges to destination;
   (c) Responsible for demurrage and storage charges.
   Once the carrier is loaded then the title passes to buyer.

2. *C.I.F.* (Cost, insurance and freight)
   Here the responsibility of the seller exists till the goods get to the port of destination and the selling price includes ocean freight and insurance upto the point of destination.

3. *C & F* (Cost and freight)
   Here responsibilities are similar to the ones laid down under CIF but in this case the buyer is to pay for insurance.

4. *F.A.S.* (Free alongside)
   Seller is to pay all costs till the goods are delivered to the wharf alongside the carrying vessel.

Rail Transport

Indian railway network is Asia's largest and world's fourth largest system. It has a tradition of competence built over the last 125 years. There has been a continuous modernising of the track, locomotives, signalling equipment and other devices for monitoring and control. Railways in India are, at present, the largest agency in freight transportation in terms of the tonnage-miles. However, over the years the percentage of freight traffic handled by road has steadily increased. Railways are ideally suited for long distance—over 500 kilometres freight transportation within the country. Movement of heavy materials such as coal, steel, raw materials, ores, etc., are most economically handled by Railways. As a national organisation, Railways have certain social obligations. Thus, they give priority to movement of foodgrains, coal, raw materials and finish products for steel plants and so on. They have therefore evolved a freight structure which will ensure subsidised transportation of these essential commodities. "The freight rate structure of Indian railways is the lowest in the world as they have to be based on social obligations and not on economic considerations."* This has made possible a shift in freight traffic to road, particularly in the case of high rated commodities. It is imperative for the materials manager to avail of the various need-based transport packages that have been developed by railways, which are discussed below:

Quick transport service. A scheme has been introduced to overcome the consumers' apprehension about undue delay in the transport of goods by introducing a guaranteed Quick Transit Service on payment of a small surcharge which is refundable in case consignments do not reach their destination within the target time. The service so far has been introduced on the main trunk routes connecting most of the big commercial centres. This is a continuing process and more and more commercial centres are being connected through this service.

Container service. To compete with the door-to-door service of road transport, particularly for smalls and high-rated commodities, the Railways introduced a scheme of road-cum-rail transport in containers of five tonnes capacity, ensuring door-to-door delivery as catered for by road transport. Here, again, the service is relatively on a small scale covering only a very few commercial centres and is being progressively extended to other commercial centres.

Freight forwarder scheme. One of thc biggest disadvantages with the consumer desirous of transporting smalls by rail has been his apprehension regarding the wagons loaded with smalls being re-packed at the various goods sheds. This has been a necessity, as the wagon loaded with smalls does not necessarily have all goods for the same destination. The freight forwarder scheme involves a collection agency which ensures that a single wagon is loaded with smalls of various consumers for a single destination, thus eliminating en-route re-packing, with possible loss and damage. This scheme again is on a small scale and is being developed.

---

* Speech of Railway Minister of Government of India in Calcutta on 18-9-1975 and AIR news bulletin on 19-9-1975.

Recently, a scheme has been introduced by the Railways for collection of smalls from specified points through Mobile Booking Service.

Parcel expresses. Introduction of an increased number of parcel expresses for quicker transport of perishables and other consignments by freight rates in coaching vans is receiving the constant attention of the Railways. In view of the international energy crisis and the rise in the price of petroleum products, there is a need to save petroleum-based products on a long term basis. It is important to recognise here that 88 per cent of the total consumption of high speed diesel oil in the country is accounted for by the transportation sector. The consumption of railways is as low as 12 per cent. The productivity of HSD oil on rail transport is said to be nearly six times that of road transport. Therefore at a national level there is a need to avoid wasteful competition between rail and road transport so that each can concentrate on traffic for which it is best suited.

Road Transport

There are a large number of transport organisations in the country having a fleet of trucks. They are suited for transporting freight of the order of 5 to 15 tonnes usually over distances of 300 to 500 kms. There are many factors hindering the movement of freight by road. Some of them are octroi, interstate permits, checkposts and high price of fuel. However, this is a very important mode of transportation of material. The Government of India, for facilitating faster movement of goods transport throughout India by road, has issued over 5,000 permits on an all India basis recently.

Movement through inland waterways, pipelines and air are other modes available. They can be fruitfully used depending upon the nature of the operations, location and urgency. For instance, crude oil from the ports to the refinery is normally conveyed through pipelines.

Route selection: Rate verification and auditing. Selection of routes is very important, especially in the case of imported consignments. Sometimes it may be worthwhile to leave the choice to the suppliers so that he can utilise his ship's freight carrying capacity in an economical manner. However, under FOB terms normally the buyer has the right to select the route. The supplier is requested by the buyer to select the route that is most economical. A description of various terms of contract with regard to imports is available in Exhibit 28.1. Detailed procedures on imports are available in the chapter on international buying.

The rates payable depend upon the volume of freight handled, nature of the materials, discount aspects, space requirements, distance to be covered, special services required such as stopovers and Governmental regulations. Terms, such as FOB, CIF also influence the final rate per tonne. It is essential that continuous monitoring of rates is done so that prompt payments are made. This will minimise costly penalties and demurrages. These days demurrage charges are prohibitively high so as to ensure that wagons are not utilised as "warehouses". In view of the general wagons shortage in the country, it is important that maximum utilisation must be made.

Lost shipments and claims. When a vast amount of freight is transported, it is but natural that some consignments will be lost or damaged. It is essential that a claim must be immediately lodged or attempts must be made to trace the consignment. Proper filing of claims will avoid wasteful correspondence. Usually the consignments are identified and checked at the central receiving section of the plant. It is beneficial to have a team at that point specifically to handle claims. The function will be to collect all the supporting documents such as the railway receipt, open delivery certificates and other relevant documents so that a complete claim could be properly lodged. A detailed discussion on "claims management" is available in Chapter 19 on Insurance Buying.

Linear Programming techniques can be used so as to minimise the transportation cost. This is described in detail in Chapter 32.

Thus we have seen the importance of transportation in relation to managing materials, the alternatives available, their merits and demerits. Every organisation must carefully analyse these aspects in relation to their specific requirements so that a judicious choice is made. Some organisations have a courier service to transport some important documents and materials. At times this is extended to inter-firm documents also. However, this is not legal and hence should be avoided.

Part IV
INVENTORY MANAGEMENT AND/OR TECHNIQUES

29

Inventory Management in India

INTRODUCTION

In financial parlance, inventory is defined as the sum of the value of raw materials, fuels and lubricants, spare parts, maintenance consumables, semi-processed materials and finished goods stock at any given point of time. The operational definition of inventory would be: the amount of raw materials, fuel and lubricants, spare parts and semi-processed material to be stocked for the smooth running of the plant. Since these resources are idle when kept in the stores, inventory is defined as an idle resource of any kind having an economic value.

Inventories are maintained basically for the operational smoothness which they can effect by uncoupling successive stages of production, whereas the monetary value of inventory serves as a guide to indicate the size of the investment made to achieve this operational convenience. The materials management department is expected to provide this operational convenience with a minimum possible investment in inventories. The objectives of inventory, operational and financial, needless to say, are conflicting. The materials department is accused of both stockouts as well as large investment in inventories. The solution lies in exercising a selective inventory control and application of inventory control techniques.

Inventory control has been attracting the attention of managers in India for a long time. But with the credit squeeze measures announced by the Government of India and the consideration of the recommendations of the Tandon Committee for inventories—see Exhibit 5.3, top management is deeply concerned with developing suitable norms for inventory control. In this context research has shown that the engineering-oriented public sector undertakings carry more inventory than the other categoreis.*

RAW MATERIAL, WIP, FINISHED GOODS

For control purposes, it is very essential to study the inventory in detail—raw materials, production components, work-in-progress and finished goods inventories

* *Inventory Control in Public Sector Undertakings* by P. Gopalakrishnan, *Economic Times,* 24 September, 1975.

should be segregated as the reasons for their existence and the cause for their size are different.

Raw materials and production components are purchased from outside suppliers and the reason for their existence is to uncouple the purchasing function from the production function. The size of this inventory is dependent upon factors such as internal lead time for purchase, supplier lead time, vendor relations, availability of the material, government import policy, in the case of imported material, the annual consumption of the materials (ABC classification) and the relative criticality of the material (VED classification).

Work-in-progress inventory might exist merely because of the production cycle time or could also be maintained for decoupling successive manufacturing operations. The decoupling could be employed either for implementing an incentive scheme or to enable each of the production departments to plan independently. The size of this inventory is dependent on the production cycle time, the percentage of machine utilisation, the make/buy policies of the company, and the management policy for decoupling the various stages of manufacturing.

The finished goods inventory is maintained to assure free-flowing supply to the customers and for this the marketing department insists on substantial finished goods inventory. The size also depends on the ability of the marketing department to push the products, the company's ability to stick to the delivery schedule of the client—the shelf life and the warehousing capacity.

Two factors which influence the inventories of all types are: the accuracy and details of the final forecast—all the inventories are geared for future requirements and are therefore sensitive to this factor—and the available storage space—the logical sequence to this factor is the shelf life of the items stored, a factor for consideration in the case of perishable goods.

NORMS FOR INVENTORY

The norms for inventory could be set by either the top management, or the materials management department. The top management usually sets monetary limits for investment in inventories. The materials department then has to allocate this investment to the various items and ensure the smooth operation of the company. It would be worthwhile if the inventory norms are set by the "management by objectives" concept. This concept expects the top management to set the inventory norms (limits) in consultation with the materials department. The norms thus evolved should be specific and quantified. The achievement of the targets set is the responsibility of the materials department.

In the setting up of the norms, the involvement of persons who are directly responsible for maintaining the inventories is very desirable. Other departments involved in setting the norms are finance, production, marketing and materials control. The norms of inventory should be converted to specifically spell out parameters like the number of stockouts permitted, the sales to inventory ratio and inventory to consumption ratio.

PECULIARITIES IN INDIA

All the well-known inventory control techniques have a basic assumption: "free availability of materials" as and when required in any quantity. This is however not true of Indian conditions. We operate in a seller's market for most of the materials. There is a perpetual scarcity of key raw materials and the prices fluctuate widely. However, the techniques should not be discarded but used judiciously as a broad guideline, keeping in mind their limitations.

The Indian industry tends to stress a lot on production and machine utilisation—this focus of attention in the context of scarcity of materials leads to hoarding of stocks. The inflation that prevails in the country, prompts hoarding.

In view of the shortage of foreign exchange, strict import procedures are enforced. The time taken for import clearance is high and the inventory of imported materials is therefore usually very high.

In its initial stage of industrial development, the country relied solely on foreign collaboration. Machinery and spare parts were readily imported. The lack of technological knowledge made the country rely on the collaborators for the estimation of the requirements of spare parts. This stock of spares is abnormally high. It is extremely evident that these spares are useless and should be written off; however, for financial reasons they continue to exist on inventory records.

Inventory control systems are built on the foundation laid by materials management techniques such as ABC-VED analysis, standardisation and codification. Setting up an inventory control system without the pre-use of these techniques is like building a castle in the air. The importance of ABC-VED analysis is brought out in the light of the inability to measure precise costs on the basis of which inventory levels could be set. Standardisation and Codification help in variety reduction and make inventory control purposeful.

Inventories, built to act as a cushion between supply and demand, serve the following needs: it is sufficient to take care of the requirements of demand till the next supply arrives, it is sufficient to take care of probable delays in supply as well as probable variations in demand.

The problems that need to be tackeld are: the determination of the level of inventory for placing a replenishment order, the quantity to be ordered, the amount of delay in supplies and the amount of variations in demand which the inventory should be able to withstand. The problems can be resolved by the cost implications. Costs which are relevant for consideration are discussed in the following paragraphs.

RELEVANT COSTS

Basically there are four costs relevant for consideration in developing an inventory model. These are: (1) the cost of placing a replenishment order, (2) the cost of carrying inventory, (3) the cost of understocking, and (4) the cost of overstocking. The cost of ordering and the inventory carrying cost are viewed as the supply side costs and help in the determination of the quantity to be ordered for each replenishment. The understocking and overstocking costs are viewed as the demand side costs and help in the determination of the amount of variations in demand and the delay in supplies which the inventory should withstand.

Cost of Ordering

Every time an order is placed for stock replenishment, certain costs are involved, and, for most practical purposes, it can be assumed that the cost per order is constant. The ordering cost—C_o—may very, dependent upon the type of items: raw material like steel against production components like casting. However, it is assumed that an estimate C_o can be obtained for a given range of items. This cost of ordering, C_o includes:

1. Paper work costs, typing and despatching an order.
2. Follow-up costs—the follow-up required to ensure timely supplies—includes the travel cost for purchase follow-up, telephone, telex and postal bills.
3. Costs involved in receiving the order, inspection, checking and handing to the stores.
4. Any set up cost of machines if charged by the supplier, either directly indicated in quotations or assessed through quotations for various quantities.
5. The salaries and wages to the purchase department. This is relevant for consideration if the purchasing function is carried out at the same level with the existing staff. If the level of purchasing activity decreases significantly, obviously a proportional amount of personnel will be transferred to other departments.

If the level of purchasing increases, the extra load will be tackled by paying overtime to existing staff or by recruiting new personnel. This additional cost can be viewed as the marginal cost of orders. The ordering cost in a typical Indian firm is around Rs 100 per order, but experience shows that this cost varies considerably depending upon the efficiency of the purchasing department.

Some firms operate on the basis of "receipt cost" particularly when dealing with staggered deliveries. The mathematical models can be suitably modified to get the "economic receipt quantity" instead of "economic order quantity".

Cost of Inventory Carrying

This cost C_c is measured as a percentage of the unit cost of the item. This measure, therefore, gives a basis for estimating what it actually costs a firm to carry stock. This cost includes: (1) interest on capital, (2) insurance and tax charges, (3) storage costs—any labour, excluding handling of receipts of new orders—the costs of provision of storage area and facilities like bins, racks, etc., (4) allowance for deterioration or spoilage, (5) salaries of stores staff, and (6) obsolescence. The inventory carrying cost varies and in a typical Indian industry is about 30 per cent. A major portion of this is accounted for by the interest on capital which depends on the fiscal policies of the government. A few firms differentiate the costs as fixed and variable. In the analysis of and use of mathematical formula, only the variable costs of ordering should be considered as the fixed costs will be constant irrespective of the number of orders placed or the inventory carried.

Understocking Cost

This cost, K_u, is the cost incurred when an item is out of stock. It includes the cost of lost production during the period of stock out and the extra cost per unit which might have to be paid for an emergency purchase.

Overstocking Cost

This cost K_o is the inventory carrying cost (which is calculated per year) for a specific period of time. The time varies in different contexts— it could be the lead time of procurement or the entire life-time of a machine. In the case of one-time purchases, overstocking cost would be: Purchase Price – Scrap Price.

Based on the above costs several scientific models have been developed and these are discussed separately for consumables and for spares in the following chapters. The need for using these models in India have been stressed by various committees. For example, the report of the Committee* on Public Undertakings mentions: "For proper inventory control, it is essential to adopt the scientific practices and techniques that have been developed in this regard".

---

* *Fortieth Report on Materials Management in Public Undertakings,* Committee on Public Undertakings, Lok Sabha Secretariat, New Delhi, p.12.

30

Economical Ordering Quantity

Inventories, as stated earlier, are built to act as a cushion between supply and demand. The supply could be a one-time supply (static) or a continuously repetitive one (dynamic). The demand on its part could be totally known (certain), be known over a range of probable values (risks), or be totally unknown (uncertain). Thus, six different inventory models can be perceived:

	Supply	*Demand*
1.	Static	Certain
2.	Static	Risk
3.	Static	Uncertain
4.	Dynamic	Certain
5.	Dynamic	Risk
6.	Dynamic	Uncertain

Out of these six possible inventory models, the static-certain model is very obvious. The supply (or requirement) is one-time and the quantity is known. Hence allowing for the lead time, the quantity can be ordered so that it is received when needed. The dynamic-uncertain model does not exist for the very reason that a situation which is dynamic and repetitive ceases to be uncertain after a certain lapse of time. The static-uncertain model is based on regret matrix which is discussed in detail in the chapter on Purchasing under Uncertainty. The three remaining models, then, are static-risk, dynamic-certain and dynamic-risk.

The costs relevant for consideration in these three models are:

Inventory	*Under-stocking cost*	*Overstocking cost*	*Inventory carrying cost*	*Ordering cost*
Static-Risk	✓	✓	—	—
Dynamic-Certain	—	—	✓	✓
Dynamic-Risk	✓	✓	✓	✓

STATIC-RISK MODEL

This model is useful for project purchases and purchase of capital and insurance spares. Examples in this category are publication of calendars, diaries, yearbooks, buying capital spares, etc., which are one-time decisions, but the quantities can be estimated in the probability values. The costs relevant for consideration are the understocking and overstocking. It is pointless to consider the supply side costs as the purchase is not repetitive. The optimal stocking policy in this situation is to stock up to level "*S*"—*S* denotes service level—where $F(S)$, the cumulative probability of demand up to "*S*", is equal to $\frac{K_u}{K_u+K_o}$. The details of this deviation are given in Exhibit 30.1. The solution of this model—based on the probability of each value of occurrence—answers the very important question of the amount of demand to be satisfied for optimal investment in such one-time decisions.

DYNAMIC-CERTAIN (EOQ) MODEL

The costs relevant for consideration in this model are the costs of ordering and inventory carrying. Since the demand is certain, the understocking and overstocking costs become irrelevant for consideration. The best way to understand the behaviour of the cost of ordering and the inventory carrying cost is by means of an illustrative example:

Demand is certain and is 32 units per month. The ordering cost is Rs 7 per order. The unit cost of the item is Rs 50 and the inventory carrying cost is 24 per cent in this example. This means an inventory carrying charge of Rs $\frac{(50 \times 24 \times 1)}{12 \times 100}$ = 1 per unit per month. The various ordering alternatives and their monthly cost is tabulated in Exhibit 30.2.

The optimum policy as indicated by the table is for placing two orders a month. The important aspect revealed, however, by this table, is the conflicting nature of the cost of ordering and inventory carrying cost. These costs move in opposite directions at different rates. This indicates the likely existence of a point where the total cost will be minimum.

The derivation of the optimal point is shown in Exhibit 30.3. The order quantity as indicated by this derivation is called the "Economic Order Quantity" (*EOQ*). The "*EOQ*" is equal to $\sqrt{2MC_o/sC_c}$ where *M* is the annual demand of the item, and *s*, the unit cost of the item, C_o the ordering cost and C_c the carrying cost.

COST-SENSITIVITY ANALYSIS

The economic order quantity is given by $\sqrt{2MC_o/sC_c}$, the values of C_c and C_o are "estimated" as average for all the items in the stores. The averages are used for convenience in designing an inventory model. It can be proved that these

EXHIBIT 30.1

OPTIMAL LEVEL OF STOCKING WHEN THE DEMAND IS UNDER RISK

Let the demand be distributed as per the probability density function $F(x)$. Let S be the stock on hand.

The expected cost of understocking

$$X = \int_S^{\infty} K_u (x - s) F(x)\, dx .$$

The expected cost of overstocking

$$= \int_0^{S} K_o (s - x) F(x)\, dx .$$

The total cost

$$G = \int_S^{\infty} K_u (x - s) F(x)\, dx + \int_0^{S} K_o (s - x) F(x)\, dx .$$

For this total cost to be minimum

$$\frac{dc}{dS} = 0 = K_u \int_S^{\infty} - F(x)\, dx + K_o \int_0^{S} F(x)\, dx$$

or

$$K_o \int_0^{S} F(x)\, dx = K_u \int_S^{\infty} F(x)\, dx$$

$$K_o[F(s)] = K_u \left[1 - \int_0^{S} F(x)\, dx \right] = K_u[1 - F(S)]$$

$$K_o F(S) = K_u - K_u F(S)$$

or

$$F(S) = \frac{K_u}{K_u + K_o}$$

$F(S)$ is also the service level.

estimates are useful for practical consideration. This is done by carrying out a cost sensitivity analysis.

Let us suppose that for a particular item, the precise cost of ordering is $W_1\ C_o$, where C_o is an average estimate of ordering cost and the precise inventory carrying cost is $W_2\ C_c$, where C_c is the average estimate of inventory carrying

EXHIBIT 30.2

Number of orders per month	*Average inventory*	*Carrying cost*	*Ordering cost*	*Total monthly cost*
One order	16	16	7	23
Two orders	8	8	14	22
Four orders	4	4	28	32

EXHIBIT 30.3
DERIVATION OF EOQ

Let the Annual Demand be M (deterministic). Let C_o be the cost of ordering, C_c, the inventory carrying cost, and s, the unit price of an item. The total annual ordering cost when the orders are placed in quantity Q is

$$\frac{M}{Q}C_o.$$

The average inventory carried is $Q/2$ and, therefore, the annual inventory carrying cost is = $Q/2,\ sC_c$.

The total annual cost is

$$= \frac{M}{Q}C_o + \frac{Q}{2}sC_c$$

The costs to be minimum

$$\frac{dc}{dQ} = 0 = -\frac{M}{Q^2}C_o + \frac{1}{2}sC_c$$

or

$$\frac{M}{Q^2}C_o = \frac{1}{2}sC_c$$

$$Q^2 = \frac{2MC_o}{sC_c}$$

or

$$Q\,(\text{optimum}) = EOQ = \sqrt{\frac{2MC_o}{sC_c}}$$

Graphic Derivation of EOQ

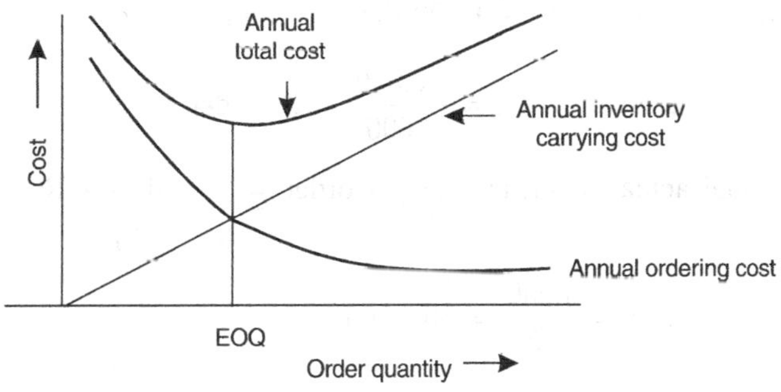

Note that the total annual cost curve is flat near *EOQ*. Refer also to cost sensitivity analysis discussed earlier. This means that rounding off of ordering quantity does not adversely affect the total inventory cost.

cost. The table below indicates the excess operating cost incurred for various values of W_1 and W_2 (see Exhibit 30.4).

EXHIBIT 30.4

Effect of W_1 and W_2 on total inventory costs

W_2 \ W_1	0.5	0.75	1.00	1.25	1.50	2.00
0.5	0	2%	6%	11%	15%	25%
0.75	2%	0	1	3	5	12
1.00	6%	1	0	1	2	6
1.25	11%	3	1	0	3	5
1.50	15%	5	2	3	0	1
2.00	25%	12	6	5	1	0

To illustrate the above table, consider an item with an annual demand of 20,000 units. The estimated cost of ordering is Rs 100 and the estimated inventory carrying cost is 20 per cent. The unit price of this item is Rs 20. Using the estimates C_o and C_c, the order quantity will be

$$\sqrt{\frac{2MC_o}{sC_c}} = \sqrt{\frac{2 \times 20{,}000 \times 100}{20 \times 0.2}}$$

$$= 1{,}000 \text{ units.}$$

If the actual cost of ordering for this item was Rs 200 ($W_1 = 2.0$) and the actual inventory carrying cost was 10 per cent ($W_2 = 0.5$) then naturally the Economic Order Quantity changes. The correct ordering quantity should have been

$$\sqrt{\frac{2M(2C_o)}{s(C_c 0.5)}} = \sqrt{\frac{2 \times 20{,}000 \times 200}{2 \times 0.10}}$$

$$= 2{,}000 \text{ units.}$$

But if the order quantity is 1,000, let us examine the consequences.

$$\text{Number of orders in a year} = \frac{20{,}000}{1{,}000} = 20 \text{ orders}$$

$$\text{Total annual actual ordering cost per order} = 20 \text{ orders} \times \text{Rs } 200$$

$$= \text{Rs } 4{,}000$$

$$\text{Average inventory} = \frac{1{,}000}{2} = 500 \text{ units}$$

$$\text{Annual inventory carrying cost} = 500 \times 20 \times \frac{10}{100}$$

$$= \text{Rs } 1{,}000$$

Therefore,

Annual inventory cost using estimates = Rs 4,000 + Rs 1,000

= Rs 5,000.

If the order quantity is in terms of EOQ then

$$\text{Number of orders placed in a year} = \frac{20{,}000}{2{,}000}$$

Annual ordering cost = 10 × Rs 200 = Rs 2,000

$$\text{Average inventory} = \frac{2{,}000}{2} = 1{,}000$$

Annual inventory carrying cost = 1,000 × Rs 20 × 10%

= Rs 2,000

Therefore,

Annual total inventory cost = Rs 2,000 + Rs 2,000

= Rs 4,000.

This means that by using average estimates, the total inventory cost incurred for the item is Rs 5,000. It would have been Rs 4,000, with correct values of C_o and C_c. This means a 25 per cent = (5,000 – 4,000)/4,000 increase only in total inventory cost. This entry of 25 per cent is made against W_1 and 2.0 and W_2 = 0.5.

Proceeding in the same manner for various values of W_1 and W_2, the table discussed above can be drawn.

In industry, there is always a tendency to discard *EOQ* as an "abominable snowman"' saying that the cost estimates can never be exact and could vary from one item to another depending upon the purchasing efforts.

The cost-sensitivity analysis table indicates that even for the extreme situations, i.e., one hundred per cent increase in ordering cost and 50 per cent decrease in carrying cost, the annual operating cost, which includes ordering plus carrying, is only 25 per cent above the optimal cost. As such, it should be possible, at least to start with, to apply the *EOQ* concept instead of waiting for precise cost figures. This analysis also justifies the treatment of ordering and carrying costs as constant for all the stores items.

IMPORTANCE OF *EOQ*

In a real life situation, demand is never deterministic. Even if demand remains known over a range of probable values, the starting point for a design of an inventory system is "*EOQ*" even though by itself it is not an inventory system.

Another important feature of "*EOQ*" is its justifying the conventional ABC and also elaborating it. This is shown below:

$$EOQ = \sqrt{\frac{2MC_o}{sC_c}}$$

Then

$$\text{Annual ordering cost} = \frac{MC_o}{EOQ}$$

$$\text{Annual inventory carrying cost} = \frac{EOQ}{2} sC_c$$

$$\text{Total inventory cost per annum} = \frac{M}{EOQ} C_o + \frac{EOQ}{2} sC_c$$

Substituting for EOQ,

$$\text{Total annual inventory cost} = \sqrt{2MsC_oC_c}$$

$$= \sqrt{2C_oC_c}\sqrt{Ms}$$

$$= \text{const.} \sqrt{Ms}$$

Since C_o C_c once estimated are constants, the total annual inventory cost is proportional to the square root of annual consumption. The illustrative example shown below indicates the savings in inventory cost by adopting selective inventory control and also that (the number of orders placed in a year) should be made proportional to the square root of annual consumption.

Example. A company purchases three items A, B and C. Their annual demand and unit prices are tabulated below:

EXHIBIT 30.5

Item	*Annual demand*	*Unit price* Rs
A	1,20,000	3
B	80,000	2
C	600	96

The company followed an ad hoc policy of placing one order per month for each of the items, totalling 36 orders per year. The annual inventory investment thus incurred is tabulated below:

EXHIBIT 30.6

Item	*Average inventory in number of units*	*Average inventory investments*
A	5,000	15,000
B	3,333	6,666
C	25	2,400
Total inventory investment		24,066

By selective inventory control, the 36 orders that are placed in a year would be distributed between A, B and C as per the square root of their annual consumption. This exercise is useful in the absence of the cost of ordering and inventory carrying.

EXHIBIT 30.7

Items	*Annual demand*	*Unit price* (Rs)	*Annual consumption* (Rs)	$\sqrt{\textit{Annual consumption}}$
A	1,20,000	3	3,60,000	600
B	80,000	2	1,60,000	400
C	600	96	57,600	240
				1,240

Number of orders placed for each item is worked out as follows:
Total number of orders placed = 36

$$= \text{constant } [\sqrt{\text{annual consumptioon of A}} + \sqrt{\text{annual consumption of B}}$$

$$+ \sqrt{\text{annual consumption of C}}\ ]$$

Therefore,

$$\text{Constant} = \frac{36}{600 + 400 + 240} = \frac{1}{34.44}$$

In a general case, the formula for determining the optimum number of orders for an item is

$$\text{Number of orders} = K\sqrt{\text{Consumption}},$$

Where

$$K = \frac{\text{total order}}{\sqrt{\text{total consumption}}}$$

The ordering pattern and the inventory investment would be as follows:

EXHIBIT 30.8

Item	$\sqrt{\textit{Annual consumption}}$	*Number of orders in a year*	*Average inventory*	*Inventory Investment*
A	600	17.4	3,444	10,332
B	400	11.6	3,444	6,888
C	240	7.0	42	4,133
		36.0		21,353

The saving in inventory investment is 24,066 – 21,353 = Rs 2,713. This capital is released from inventory even without going into details of estimation of ordering and inventory carrying costs.

If ordering and inventory carrying costs are estimated, it is very easy to tabulate "*EOQ*" value in Rupees for various values of Annual Consumption. As already indicated, the "*EOQ*" value is proportional to the square root of annual consumption.

Example. For C_o = Rs 90 per order C_c = 20%, the table below indicates the "*EOQ*" values for various values of annual consumption.

$$EOQ \times s = \sqrt{\frac{2C_oM}{sC_c}}\, s = \sqrt{\frac{2C_o}{C_c}Ms}$$

In the example,

$$\sqrt{\frac{2C_o}{C_c}} = \sqrt{\frac{2\times 90}{0.20}} = \sqrt{900} = 30$$

This is taken as a constant for a firm and given the annual consumption value, the *EOQ* value can be worked out as in the table below:

EXHIBIT 30.9

Annual consumption	*EOQs*	*Optimal number of orders a year* $= \frac{Ms}{EOQs}$
1,00,000	9,480	10.5
40,000	6,000	6.6
10,000	3,000	3.3
8,100	2,700	3.0
4,900	2,100	2.33
1,600	1,200	1.33
900	900	1.00
100	300	0.33
50	210	0.23
25	150	0.17

ECONOMIC PURCHASE QUANTITY

From the table above it is imperative to conclude that for operational convenience it would be worthwhile rounding off the number of orders. The other reasons for deviating from *EOQ* to Economic Purchase Quantity are given below:

1. Ordering to the nearest trade quantity or packing quantity. In this context, it must be mentioned that EOQ for items with low M (annual demand) and high s (unit price) will be small; the suppliers will find it uneconomic to produce and hence not quote for this quantity.
2. The order quantity could be altered to get benefits of transport—getting a truck-load or a wagon-load.
3. In the case of perishable goods the quantity has to be adjusted to shelf life of the item.
4. When prices fluctuate widely, a different approach (dynamic-programming) has to be used to determine the quantity to be purchased.
5. Seasonal supply factors force organisations to buy in large quantities.
6. In the case of imported items companies are forced to buy at least six months' requirements at a time.
7. Discounts offered for bulk purchases suggest larger quantities than EOQ. This aspect of accepting a discount is illustrated by an example.

Example. In a leading biscuit manufacturing organisation, the annual demand for corrugated boxes is 20,000. The cost of placing an order is Rs 100 and the inventory carrying cost is 20 per cent. The price per box is Rs 10. The supplier offers 1 per cent discount if 4,000 corrugated boxes or more are purchased and 3½ per cent discount if 10,000 units or more are purchased. What should be the ordering quantity and would you accept the discount?

Solution

For standard rate of Rs 10 unit,

$$EOQ = \sqrt{\frac{2C_oM}{sC_c}}, \quad M = 20{,}000 \quad C_o = 100 \quad s = 10 \quad C_c = 20\%$$

$$EOQ = \sqrt{\frac{2(20{,}000)\,100}{10(0.20)}} = 1{,}440 \text{ units.}$$

$$\text{No. of orders placed in a year} = \frac{20{,}000}{1{,}440}$$

$$= 14.4 \text{ orders.}$$

The average inventory is

$$\frac{1{,}440}{2} = 720 \text{ units.}$$

Let us now examine the consequences of discount of 1 per cent for buying 4,000 or more units at a time.

If the total saving from discount offer—i.e., amount saved (price discount) in annual purchase + amount saved in ordering—exceeds the extra cost incurred is the inventory carrying cost, then accept the discount; otherwise, reject it. In the example, savings in annual purchase and ordering can be worked as follows:

$$\frac{1}{100} \times 20{,}000 \times 10 + \left(14.4 - \frac{20{,}000}{4{,}000}\right)100 = 2{,}000 + 940 = \text{Rs } 2{,}940.$$

The extra inventory carrying cost is $(4{,}000 - 1{,}440) \times \frac{1}{2} \times 10 \times 0.20 = \text{Rs } 2{,}560$.

It is assumed that inventory is valued at a standard rate of 10/unit. Since the savings are greater than the additional inventory, accept the discount.

Let us now examine the consequences of buying in quantity of 10,000.

The savings can be determined as follows:

$$\frac{7}{2} \times \frac{1}{100} \times 20{,}000 \times 10 + \left(14.4 - \frac{20{,}000}{10{,}000}\right) \times 100$$

$$= 7{,}000 + 1{,}240$$

$$= \text{Rs } 8{,}240.$$

Additional inventory costs

$$= (10{,}000 - 1{,}440)\ \frac{1}{2} \times 10 \times 0.20$$

$$= \text{Rs } 8{,}560.$$

Since additional costs are greater than the savings, reject the discount offer. The order quantity should therefore be 4,000 units.

DYNAMIC RISK

This is the most practical inventory model. This is the situation faced in industries. The model is therefore used as an approach to building up of inventory systems. These inventory systems are discussed in detail in the next chapter.

Practical Inventory Systems

SYSTEM DESIGN

In this chapter, we will discuss the dynamic risk situation which exists in practice. Here we have to consider inventory carrying costs, ordering costs, understocking and overstocking costs. The design of an inventory system starts with the determination of *EOQ*. While *EOQ* indicates the quantity to be ordered, it, however, does not indicate when to place the order. This, known as the re-order point, is determined by the consumption during the lead time. This is depicted in the diagram below with no buffer stock as demand as well as lead time are certain and known.

EXHIBIT 31.1

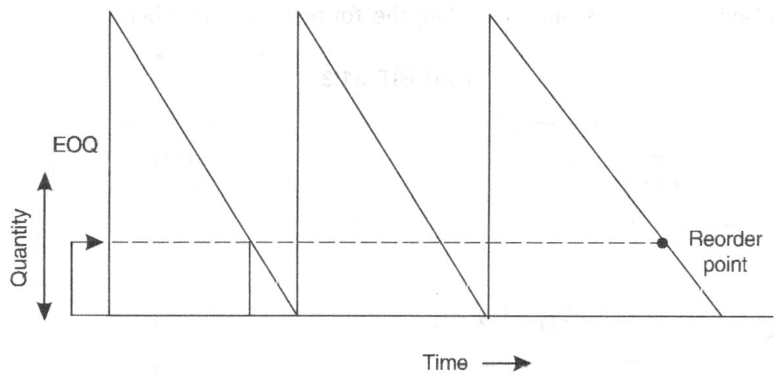

In practice, however, it is quite likely that the lead-time consumption value is greater than *EOQ*. Let us consider a situation where lead time is 67 days and *EOQ* works out to 30 days' consumption. In such a case, the inventory system should be started with a purchase of lead-time consumption value and then brought down to repetition of *EOQ* cycles. Here we start with an initial stock equal to lead time consumption. At this stage an order for *EOQ*—O_1—is placed which is received after 67 days (R^1). At intervals of every 30 days—*EOQ* expressed in terms of number of days consumption, place an order for *EOQ*. This system is depiced in Exhibit 31.2.

EXHIBIT 31.2

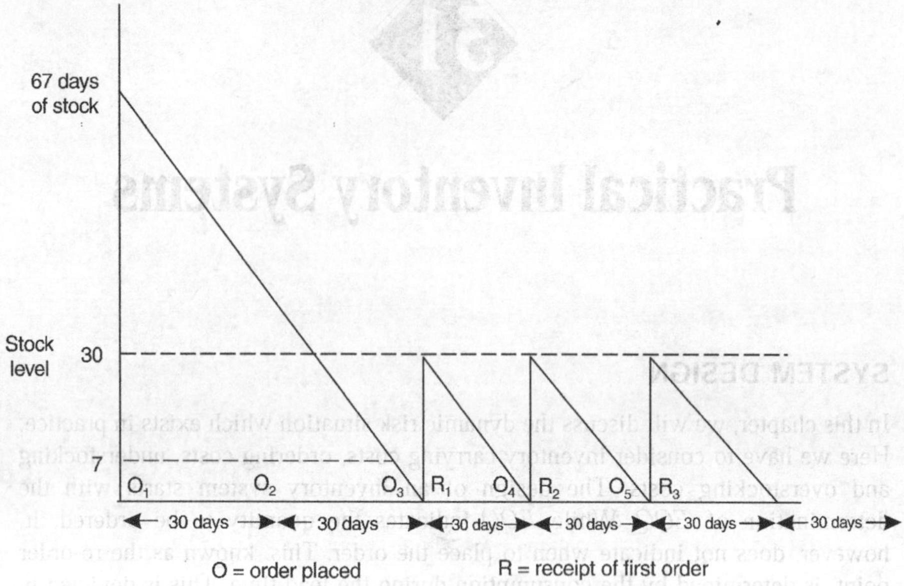

Gap between the receipts will be one month.

Gap between the ordering will be one month.

It can be clearly noted from the diagram that orders are triggered off whenever the quantity on hand plus on order is equal to the lead time consumption. A generalised inventory system can then the formed as shown below:

EXHIBIT 31.3

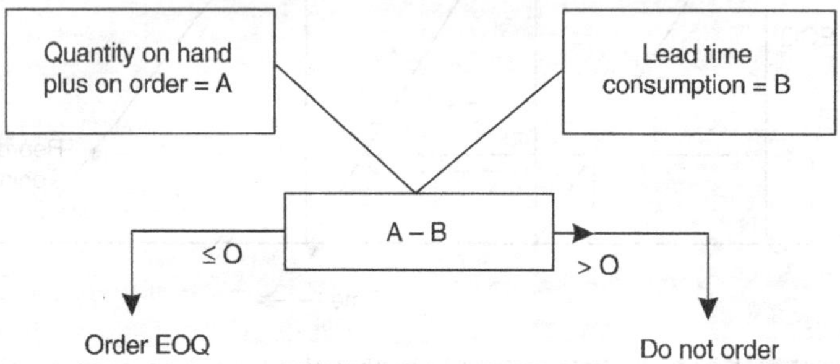

SAFETY STOCK

In real life situations, two factors play an important role in the design of a system. The first factor is that demand is never certain and is at best distributed over a set of probable values. Secondly, the lead time is never fixed but varies over a certain range of days. Hence, we have taken precautions for the above two factors. A company can tackle this situation by ad hoc policies, such as treating demand as 10–20 per cent more than the average demand or maintaining inventories' levels at 10–20 per cent more than the previous period.

The scientific design of an inventory system is based on the optimisation of overstocking and understocking cost. The stock level, as already explained in the previous chapter, is set by $F(S) = K_u/(K_u + K_o)$. The distribution from which $F(S)$ is read off is the normal distribution given in Exhibit 31.6. It is usually seen that the frequency distribution of consumption for most of the regular items is a normal distribution which is uniquely determined by the mean and standard deviation—Exhibit 31.6. The calculation of mean and standard deviation is given in Exhibit 31.5. This approach could be used to see the extent to which it is economical to guard against demand variations. The reserve stock to take care of such demand variations is obtained through multiplying the standard deviation of the demand by the factor read off from the normal tables corresponding to the given service level. The lead-time variations are best taken care of by summarising the historic data on lead-time and studying the lead-time distribution pattern. A decision can then be taken to carry inventory to guard against delays and the level of inventory depends on the probability of delay and the maximum delay that can occur. The 'safety stock' to act as a cushion for the lead-time fluctuations is determined by the average demand during maximum delay multiplied by the probability of delay. The safety stock is then added to the reserve stock in order to get the cushion which in common parlance is called the 'safety'. The two inventory systems which utilise these scientific ways of determining inventory levels are *Q*-system, the quantity ordered with each order is fixed (*EOQ*-based on average demand). The number of orders to be placed in a year is made, dependent on variations in demand and lead-time. In the *P*-system, the number of orders to be placed is fixed depending on the review time, time between two successive orders. The quantity to be ordered is made, dependent on the variations in demand and lead-time. These two systems are explained in great detail with an example.

Q-SYSTEM

The quantity ordered is *EOQ*. The reorder point is determined by three components:

1. The average demand during the average lead-time which in scientific terminology is known as buffer stock.
2. The variations in demand during average lead-time known as reserve stock which depends on the service level.
3. The average demand during delivery delays is known as safety stock which is obtained through multiplying average demand for maximum delay and the probability of delay. A company should always carry safety stock. Any depletion in safety stock should be made good immediately with the next order. The sum of the above three components is the reorder point and an order for *EOQ* is placed as soon as physical stock reaches the reorder point. In this case the *EOQ* expressed as number of days consumption is greater than average lead-time consumption. This system is shown in Exhibit 31.4.

EXHIBIT 31.4

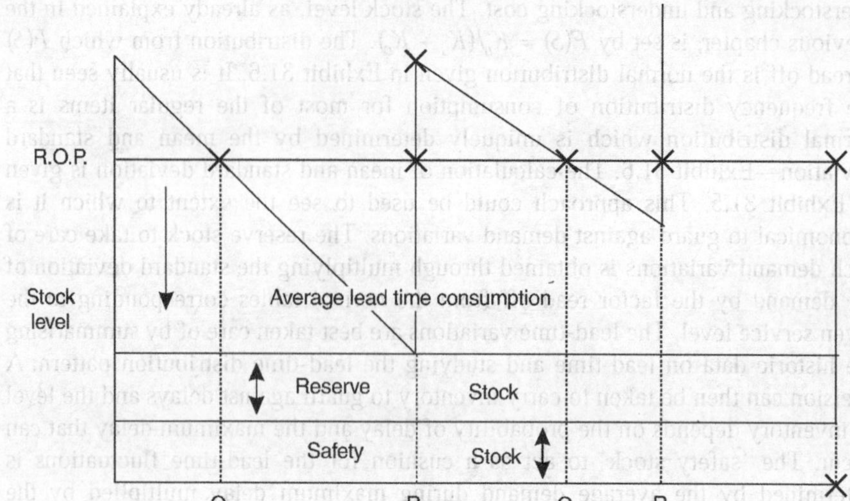

EXHIBIT 31.5
STEP TO GET STANDARD DEVIATION

In a set of observations, the measure of fluctuation is provided by the standard deviation. This is defined as the root mean square of the deviations of individual observations from the mean and is explained below by an example.

Step 1: Get the data on monthly consumption figures for a period of 24 months as follows:

26, 26, 25, 26, 25, 28, 32, 33, 26, 35, 30, 27, 36, 23, 32, 31, 36, 27, 31, 31, 39, 32, 25 and 38

Step 2: Total the values and divide by the total number of observations. This arithmetic mean = 30.

Step 3: Get the differences of individual observations from the mean as follows:

– 4, – 4, – 5, – 4, – 5, – 2, 2, 3, – 4, 5, 0, – 3, 6, – 7, 2, 1, 6, – 3, 1, 1, 9, 2, – 5 and 8.

Step 4: Square these differences:

16, 16, 25, 16, 25, 4, 4, 9, 16, 25, 0, 9, 36, 49, 4, 1, 36, 9, 1, 1, 81, 4, 25 and 64.

Step 5: Add these squares. This value is 476.

Step 6: Divide the total in step 5 by one less than the number of observations and extract the square root. This is the standard deviation.

$$sd = \sqrt{\frac{476}{23}} = 4.6.$$

In the case of regularly consumable items, the frequency distribution is normal and usually the future consumption values will fall in an interval of mean ± 35.0 namely 30 ± 13.8 or between 26.2 to 43.8.

EXHIBIT 31.6
AREAS UNDER THE NORMAL CURVE

Proportions of observations lying above a distance of *ts* from the mean
sd = standard deviation

t	*Proportion*	*t*	*Proportion*	*t*	*Proportion*
–4.0	0.99997	–1.3	0.9032	1.4	0.0808
–3.9	0.99995	–1.2	0.8849	1.5	0.0668
–3.8	0.99993	–1.1	0.8643	1.6	0.0548
–3.7	0.99989	–1.0	0.8413	1.7	0.0446
–3.6	0.99984	–0.9	0.8159	1.8	0.0359
–3.5	0.99977	–0.8	0.7881	1.9	0.0287
–3.4	0.99966	–0.7	0.7580	2.0	0.0228
–3.3	0.00052	–0.6	0.7257	2.1	0.0179
–3.2	0.99931	–0.5	0.6915	2.2	0.0139
–3.1	0.9990	–0.4	0.6554	2.3	0.0107
–3.0	0.9987	–0.3	0.6179	2.4	0.0082
–2.9	0.9981	–0.2	0.5793	2.5	0.0062
–2.8	0.9974	–0.1	0.5398	2.6	0.0047
–2.7	0.9963	0.0	0.5000	2.7	0.0035
–2.6	0.9953	0.1	0.4602	2.8	0.0026
–2.5	0.9988	0.2	0.4207	2.9	0.0019
–2.4	0.9918	0.3	0.3821	3.0	0.0013
–2.3	0.9893	0.4	0.3446	3.1	0.0010
–2.2	0.9861	0.5	0.3085	3.2	0.00069
–2.1	0.9821	0.6	0.2743	3.3	0.00048
–2.0	0.9772	0.7	0.2420	3.4	0.00034
–1.9	0.9713	0.8	0.2119	3.5	0.00023
–1.8	0.9641	0.9	0.1841	3.6	0.00016
–1.7	0.9554	1.0	0.1587	3.7	0.00011
–1.6	0.9452	1.1	0.1357	3.8	0.00007
–1.5	0.9332	1.2	0.1151	3.9	0.00005
–1.4	0.9191	1.3	0.0968	4.0	0.00003

We note the following from this table:
Mean ±1 *Sd* includes 68 per cent of values
Mean ± 2 *Sd* includes 95 per cent of values
Mean ± 3 *Sd* includes practically all values.

Time

When stock depletes to the reorder point—i.e., at O_1,—order *EOQ*. Assuming that this order arrives at average lead-time—i.e., at R_1—the position is as shown in the diagram. At O_2 now place an order for *EOQ*. We have considered a case, where demand has decreased considerably and the lead-time is also reduced leading to an increase in stock level at the point R_2. This kind of fluctuation in demand and lead-time will level off over a period of time. However, if there is a significant change in demand pattern, then *EOQ* has to be recalculated. Similarly if there is a significant change in the lead-time pattern, the buffer stock is to be redefined. It will be appreciated that such continuous monitoring is necessary for proper inventory control.

The Q-system requires a continuous review of the inventory. It requires the maintenance of a kardex system for stocks and timely entries of receipts and issues. The kardex is required to be checked with every entry to see whether a fresh order is to be placed. In case the average lead-time is greater than *EOQ* consumption, i.e., the kardex should also have a provision to indicate the quantity on order. This system is useful for high consumption of "A" class items. The items in this case are few and it is worthwhile to have continuous scrutiny of inventory. Continuous monitoring through computer is necessary for this as orders for replacement are placed as soon as physical stock reaches the reorder point.

P-SYSTEM

In this system, the inventory review period is fixed by estimating *EOQ/M.* A desired inventory level—*DIL*—is then calculated as the sum of average demand during average lead-time plus review time (buffer stock), variations in demand during average lead-time plus review time (reserve stock) and average demand during delays in supply (safety stock).

The inventory on hand plus outstanding orders is checked every time the inventory is reviewed. The quantity ordered on each review is the difference between the desired inventory level and the inventory on hand plus on order.

Since the system takes care of variations in demand during lead-time and review time, the levels on inventory in this are obviously higher than the levels in Q-System. Thus, it is not an effective inventory system for high value "A" items. The system, however, is extremely useful for "B" and "C" class items. This is because items in these classes can be grouped together on the basis of the same review time. The review is the same for items with the same annual consumption value.

$$\text{Review period} = \frac{EOQ}{M} = \frac{EOQs}{Ms} = \frac{\sqrt{\dfrac{2MC_o}{sC_c} \times ss}}{Ms}$$

$$= \frac{\sqrt{2C_o/C_c}}{\sqrt{Ms}}$$

Since the numerator is constant, the review period is inversely proportional to the square root of annual consumption. This means that items can be grouped for combined reviews based on their annual consumption. As explained in the earlier chapter on Cost Sensitivity Analysis, the investment is only 25 per cent extra when *EOQ* is doubled or halved. This implies that for items with *EOQ* varying from 1,000 to 4,000 the review period can be estimated on the basis of a standard *EOQ* of 2,000 and the items falling in a range of consumption can be grouped in one review time. This would facilitate the application of computers to regulate the system. A major advantage of this system is in the inventory control for spare parts discussed in detail in the chapter on Spare parts.

Example. Consider the following available data and let us work out the inventory under the two systems.

Annual average demand for an item is 20,000 units; Standard deviation of demand per week = 50 units; Unit price is Rs. 10; Ordering cost = Rs 100; Inventory carrying cost = 20 per cent; Average lead-time = 4 weeks; maximum delay = 3 weeks; Probability of delay = 0.31; Service level $F(S)$ = 95 per cent.

Q-system. Order quantity = *EOQ*

$$= \sqrt{\frac{2MC_o}{s \cdot C_c}}$$

$$= \sqrt{\frac{2 \cdot 20{,}000 \cdot 100}{10(0.20)}}$$

$$= 1{,}414 \text{ units}$$

Reorder point = Buffer + Reserve + Safety,

$$\text{Buffer} = \text{Average demand in lead time} = \frac{20{,}000}{52} \times 4 = 1{,}540 \text{ units,}$$

Reserve stock = Variations in demand during lead time = factor from normal tables – for 95 per cent, (standard deviation),

Service level = 1.64,

Standard deviation during lead time = $\sqrt{4} \times 50$ = 100 units

Reserve stock = 1.64 × 100 = 164 units

Safety stock = Average demand during maximum delay × probability of

$$\text{delay} = \frac{(20{,}000 \times 3)}{52} \times 0.31 = 1{,}154 \times 0.31 = 358.00$$

Reorder point = 1,540 + 164 + 358 = 2,062 units.

P-system. Review period $= \dfrac{EOQ}{M} = \dfrac{1{,}414}{20{,}000} \text{years} = \dfrac{1{,}414}{20{,}000} \times 52 \text{ weeks}$

$$= 3.7 \text{ weeks.}$$

This has to be rounded off to either 3 weeks or 4 weeks. The choice is based on cost consideration.

32

OR Techniques in Materials Management

WHAT IS OR?

"OR Techniques" have great significance in materials management functions. In some of the previous chapters we referred to various techniques such as inventory control, dynamic programming, linear programming, sampling, learning curve techniques, exponential smoothing, and so on. In this chapter, we will deal with some of the OR techniques which have been successfully used by materials managers in India.

OR is a scientific way of tackling complex problems in the management of large systems of men, machines, money and materials. The purpose is to help management in scientific decision-making. The hallmark of OR is the quantitative approach to problem-solving, incorporating measurements of factors such as risk, uncertainty, etc., so that production and comparison of alternative decisions, strategies and controls are possible.

We can broadly identify the following steps in OR approach:

1. Definition of the objectives in specific terms, understood and accepted by the decision-makers.
2. Complete description of problems in a quantitative model so that various alternatives can be generated and compared to achieve the defined objectives.
3. To analyse the behaviour of the various factors influencing the objectives.
4. To evaluate the various solutions and their accuracy.

We will discuss some of the OR techniques now.

LINEAR PROGRAMMING FOR STOCK ALLOCATION

Linear Programming (LP) technique is successfully used for optimal decisions with regard to product-mix, location and number of warehouses, choice of transportation-mix, and so on. Let us consider the application of LP in transportation decisions.

The transportation model optimises the total cost of transportation when an organisation has its plants and warehouses located at different places in the country. Let us consider that Edna Batteries have plants at X, Y and Z. Its product "torch batteries" is sent in trucks to its warehouses situated at A, B, C and D for final delivery. The shipment, production runs and storage capacities are given on the basis of truckloads per week. The costing for transportation is also on the basis of truckloads.

The plants have their working capacities and the warehouses are of different sizes depending on market demand. Exhibit 32.1 shows the capacity details of the plants and the warehouses. The transportation cost factors are worked out in Exhibit 32.2. The logistics manager of Edna batteries wonders how the supplies from the plant be allocated to the warehouses to minimise the total transportation cost.

EXHIBIT 32.1

Warehouse	*Capacity truckloads*	*Factory*	*Production truckloads*
A	16	X	48
B	20	Y	32
C	40	Z	40
D	44		
Total:	120	Total:	120

EXHIBIT 32.2

Transportation Cost per Truckload (in Units of Rs. 100)

	A	*B*	*C*	*D*
X	6	11	3.5	6
Y	2	6	5	4
Z	1.5	11	4.5	3

SOLUTION TO THE TRANSPORTATION PROBLEM

The first step is to cast the information as a matrix of source to destination. This is done in Exhibit 32.3. The marginal rows and columns indicate the capacities and availabilities. The body of the matrix contains the cost of transportation between any two points. The next step is to find an arbitrary solution meeting the restrictions. This is called the initial feasible solution. Based on this, improvements are made to reduce the total transportation cost. The initial feasible solution is obtained through the "northwest corner method". The procedure is to start the allocation on the route in the left hand topmost corner of the matrix, and continue the process of allocation matching production capacity with storage. This is shown in Exhibit 32.4. The cost of transportation is computed in Exhibit 32.5 and is equal to Rs 63,400.

EXHIBIT 32.3

Sources		Destination A	B	C	D	Production capacity
	X	6	11	3.5	6	48
	Y	2	6	5	4	32
	Z	1.5	11	4.5	3	40
	Storage Capacity	16	20	40	44	120 / 120

EXHIBIT 32.4
NORTH-WEST CORNER ALLOCATION

	A	B	C	D	Capacity
X	6 (16)	11 (20)	3.5 (12)	6	48
Y	2	6	5 (28)	4 (4)	32
Z	1.5	11	4.5	3 (40)	40
	16	20	40	44	120 / 120

EXHIBIT 32.5
VALUES OF *U* AND *V*

					U
	6 (16)	11 – (20)	3.5 + (12)	6 3.5	0
	2 – 5.5	6 + – 6.5*	5 – (28)	4 (4)	1.5
	1.5 – 5	11 – 0.5	4.5 0.5	3 (40)	0.5
V	6	11	3.5	2.5	

* leaving variable

Cj – (ui + vj)

T.C. = 6 × 16 + 11 × 20 + 3.5 × 12 + 5 × 28 + 4 × 4 + 3 × 40
= 96 + 220 + 42 + 140 + 16 + 120
= 634 × 100 = 63,400 Rs.

TEST FOR OPTIMALITY

Define a function in terms of *u*, *v* for the unit cost of transportation. *Whenever* there is an allocation, the unit cost equals the sum of *u* and *v*. The unit cost of transportation $C = u + v$.

Start with an arbitrary value of $u = 0$, and derive the set of values u and v (u's for plants and v's for destination). For factory X the value of U is taken as 0, and the sets of values of u and v are derived in Exhibit 32.5.

The value $(u + v)$ for the *empty* cells (those where allocations are not made in the matrix) is defined as the imputed economic cost of transportation ($Z = u + v$ for empty cells). The marginal cost of changing the allocation cell by cell is obtained by $(C - Z)$ for the empty cells where C is the unit cost of transportation. If any of these marginal costs are negative, then it means that savings can be effected by reallocation. These marginal cost factors are shown in Exhibit 32.5. Factory X to Warehouse B shows a saving at the current level of allocation. Now allocate the 20 truckloads to B and complete the reallocation. This second allocation is shown in Exhibit 32.6. The total cost for this reallocation shows a decreased value of Rs 50,000. Once again values of u and v are computed and the process is continued till the marginal cost values are non-negative. This process is shown in Exhibits 32.7 through 32.13.

EXHIBIT 32.6
SECOND ALLOCATION

	[6] (16)	[11]	[3.5] (32)	[6]	48
	[2]	[6] (20)	[5] (8)	[4] (4)	32
	[1.5]	[11]	[4.5]	[3] (40)	40
	[1] 16	20	40	44	

T.C. $= 6 \times 16 + 3.5 \times 32 + 6 \times 20 + 5 \times 8 + 4 \times 4 + 3 \times 40$
$= 96 + 112 + 120 + 40 + 16 + 120$
$= 504 \times 100 = 50{,}400$ Rs.

EXHIBIT 32.7
VALUES OF *U* AND *V*

					U
	[6] – (16)	[11] 6.5	[3.5] + (32)	[6] 3.5	0
	[2] + – 5.5*	[6] (20)	[5] – (8)	[4] (4)	1.5
	[1.5] – 5.0	[11] 6.0	[4.5] 0.5	[3] (40)	0.5
V	6	4.5	3.5	2.5	

* leaving variable

EXHIBIT 32.8
THIRD ALLOCATION

	6; (8)	11	3.5; (40)	6	48
	2; (8)	6; (20)	5	4; (4)	32
	1.5	11	4.5	3; (40)	40
	16	20	40	44	

T.C. = 6 × 8 + 3.5 × 40 + 2 × 8 + 6 × 20 + 4 × 4 + 3 × 40

= 48 + 140 + 16 + 120 16 + 120

= Rs. 460 × 100 = 46,000

EXHIBIT 32.9
VALUES OF *U* AND *V*

					U
	6; − (8)	11; 1	3.5; (40)	6; + − 2*	0
	2; + (8)	6; (20)	5; 5.5	4; − (4)	– 4
	1.5; 0.5	11; 6	4.5; 6	3; (40)	– 5
V	6	10	3.5	8	

EXHIBIT 32.10
FOURTH ALLOCATION

	6; (4)	11	3.5; (40)	6; (4)	48
	2; (12)	6; (20)	5	4	32
	1.5	11	4.5	3; (40)	40
	16	20	40	44	

T.C. = 6 × 4 + 3.5 × 40 + 6 × 4 + 2 × 12 + 6 × 20 + 3 × 40

= 24 + 140 + 24 + 24 + 120 + 120

= Rs 452 × 100 = Rs. 45,200

EXHIBIT 32.11
VALUES OF U AND V

					U
	[6] − (4)	[11] 1	[3.5] (40)	[6] + (4)	0
	[2] (12)	[6] (20)	[5] 5.5	[4] 2	− 4
	[1.5] + − 1.5*	[11] 4	[4.5] 4	[3] (40)	− 3
V	6	10	3.5	6	

EXHIBIT 32.12
FIFTH ALLOCATION

	[6]	[11]	[3.5] (40)	[6] (8)	(48)
	[2] (12)	[6] (20)	[5]	[4]	32
	[1.5] (4)	[11]	[4.5]	[3] (36)	40
	16	20	40	44	

EXHIBIT 32.13
VALUES OF V AND U ARE ALL NON-NEGATIVE.
HENCE THE ALLOCATION IS OPTIMUM.

					U
	[6] 1.5	[11] 3	[3.5] (40)	[6] (8)	0
	[2] (12)	[6] (20)	[5] 3.5	[4] 0	− 2
	[1.5] (4)	[11] 6	[4.5] 4	[3] (36)	− 3
V	4.5	8	3.5	6	

LEARNING CURVE FOR NEGOTIATIONS

This concept has been successfully used by firms in the determination of the price of the item with a new source. The supplier gains knowledge normally at the buyer's cost and hence this experience should be taken advantage of in the cost structure. This concept is discussed at length below:

The first step to apply the learning curve is to identify "learning factor". This is normally done by trial and error and is particularly applicable for repetitive labour-intensive jobs. Once the learning factor is known, it is possible to determine the overall effect of cost on the basis of the number of units and in this context Exhibit 32.14 is useful. It is also possible to use the learning curve for negotiating the price with the sellers. Consider the example shown below. The Purchase Manager knows that the first unit takes 100 hours for a supplier and the second 90 hours only. Then he can work out a learning factor as 90 per cent. At 90 per cent learning factor the third unit will require 81 and so on. Thus cumulative average times can be worked out as shown in Exhibit 32.14. So by collecting statistics and analysing, the learning factor appropriate to the system in question can be found.

EXHIBIT 32.14

LEARNING CURVE—UNIT VALUE (%)

Learning factor	*75*	*80*	*85*	*90*
Unit: 1	100.0	100.0	100.0	100.0
5	51.3	59.6	68.6	78.3
10	38.5	47.7	58.3	70.3
15	32.5	41.8	52.0	66.3
20	28.8	38.2	49.5	63.4
25	26.3	35.5	47.0	61.3
30	24.4	33.5	45.0	59.6
35	22.9	31.8	43.4	58.3
40	21.6	30.5	42.1	57.1
45	20.6	29.4	41.0	56.1
50	19.7	28.4	40.0	55.2

The learning curve is a mathematical formulation quantifying the relationship between productivity and experience. As a worker gains experience his productivity increases. If a plot is made of the workers' productivity and the time on the job, the graph would look as in Exhibit 32.15. The concept is particularly valid for labour-intensive jobs.

EXHIBIT 32.15

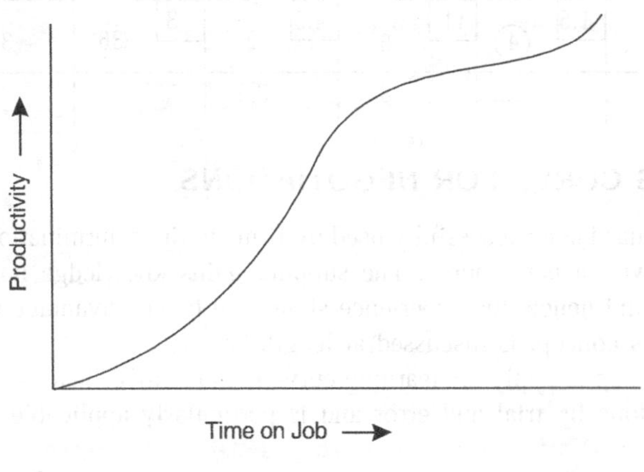

The curve rises steeply initially and then trails off, indicating a rapid decrease in time required/unit initially and then a levelling off, when productivity reaches a steady state. The learning curve evolved from the US aircraft industry, where it was found that the 4th plane required 80 per cent as much direct labour as the second, the 8th plane only 80 per cent as much as the fourth, the hundredth only 80 per cent as much as the fiftieth and so on. Due to this characteristic, the curves are also referred to as the per cent curves. The example given above would hence be an 80 per cent curves. Similarly, one could have 70 per cent, 90 per cent curves, etc.

USING AND DETERMINING THE LEARNING CURVE PARAMETER

The determination of the quantity *h*, — the learning factor—is the only area where some difficulty may be experienced. However, it has been found that in the same industry this value remains constant, i.e., the learning rate is almost the same even when there are model changes. Hence, in the US aircraft industry it was found that the 80 per cent curve fitted all models introduced and is used for determining costs for all new models which are incorporated.

The learning curve can be used for determining the labour costs of new introductions, the man-power requirements, etc. It can also be used as a central device for cost control and can be of profound implications on prices and profits.

Though the tool has achieved some standing it has not been used widely. One of the major difficulties often faced in using this technique is caused by the attitude of organised labour. Once a production line has been set, any change envisaging increase in production by using the 'learning' is resisted by the workers. Some companies were said to have solved the problem by planning the changes required before setting up the initial line. This previous knowledge of planned change helps reduce the labour's resistance to it. Further, when a company introduces a new model, some amount of 'learning' from the previous model is retained. The result is that it becomes difficult to identify the point on the curve at which the company is. If the retention is claimed equivalent to producing 50 units then for an 80 per cent curve the 100th unit would require only 80 per cent of the man-hours, but if it is at 100 units, then the man-hours go down by 20 per cent only after the next 100 units. However, this difficulty can be overcome by taking some experimental readings which would help in locating the point on the learning curve at which the force is working.

While computing the precise relationship between costs and learning, one must be alert to certain factors that can distort the picture. If costs are measured in rupees, it is necessary to eliminate inflation when observations are made over a substantial period of time. Deflation of cost figures thus becomes more important when growth rates are slow, so the doubling of trials of units requires several years.

The fact that manufacturing costs tend to follow an experience curve is not only useful for cost control and forecasting but also has a profound implication for prices and profits. In particular, it is strongly suggested that the producer of a particular product who has made the most units should have the lowest costs and highest profits.

DELPHI CONCEPT FOR FUTUROLOGY

Delphi is a fairly new technique where a set of procedures is laid down for obtaining and refining the opinions of a group of people. When the experts are asked to give their opinions, usually there is a divergence. To the decision-maker, it poses the problem of evaluation of the divergent viewpoints. Also, a group interaction has its drawbacks such as the influence exercised by dominant individuals and so on. In Delphi techniques such drawbacks are kept at a minimum.

The Delphi technique is an iterative process to obtain the most reliable consensus opinion of a group of experts. It is essentially a series of intensive interrogations of an individual, considered an an EXPERT in the relevant area of inquiry, through a number of ROUNDS of "inquiry, response and feedback". At the end of each round, the responses are analysed with statistical methods and the results are sent to the respondents for further inquiry. This is repeated until a convergence of opinions takes place. However, at each round the individual's responses are kept confidential.

In the entire process of study there is no face-to-face communication. The anonymity and secrecy of the individual's behaviour is maintained right through. Delphi technique, as a contract to an open meeting of respondents, has the following advantages: (a) the influence of dominant members and group pressure in expressing opinion is prevented, (b) the respondents are not affected by the 'band wagon effect' of a majority opinion, and (c) the respondents are free to retreat from an expressed opinion.

This technique has been widely and successfully used to forecast the distant future. The basis for the forecast is the respondent's experience and sense of awareness as an expert. This technique can be effectively used to forecast both qualitative and quantitative factors. This technique can be used to process both opinion, judgment and exact numeric factors.

Delphi procedures can be analysed under the following heads:

1. Anonymity

Anonymity of the respondents is ensured through separate and private answering of prepared questions. This means that a dominant individual cannot influence the opinions. At the same time interactions between the group members are also ensured through formal communication channels operated by the consultants or executives who control the Delphi experiment.

2. Iterative Process

The opinions and estimates of the group members are statistically analysed. For example, if the group task is to estimate a numerical quality such as forecast of materials budget after 20 years, the median values of individual estimates may be a very good indicator. True, there may be a total divergence of opinions. In such cases the results of the first round are analysed and the median and deviations are worked out. This is fed back to the respondents and they are asked to revise the estimates where appropriate, along with other members' estimates. Where members estimates are varying too much even over the interquartile range, they are asked to justify their estimates. Once again these justifications are analysed, summarised, fed back and counter-arguments are elicited. Thus it is an iterative process.

3. Group Response

Each respondent is asked to rate various questions in the order of his competence to answer them and in the order of his confidence in his answer. This helps in arriving at group response and to select a subgroup of relatively more confident individuals within the group, whose performance is better than the group as a whole. Delphi techniques are used increasingly in the areas of long range activities such as technological and social developments. It may be possible to apply this technique in materials management in areas such as new source developments, new materials, their prices, government policy on imports and lead times.

33

Role of PERT in Materials Management

One of the important problems concerning today's materials executive is planning for materials ensuring that the various stages of work are brought forward in a logical sequence with minimum cost and least delay. The Programme Evaluation Review Technique (abbreviated as PERT) is a simple but practical tool which can be used by the manager for the purpose of planning and control. PERT represents a significant step towards an integrated management system comprising variables of time, resources, technical performance and other utilities. PERT has been effectively applied in many Indian situations.

WHAT IS PERT?

PERT is one of the techniques of scientific management used in planning a project with minimum possible scheduling time and with minimum possible use of other resources. In case of uncertainties, it provides a set of principles and techniques to assist the manager. PERT pinpoints critical areas where the completion of a job on time is threatened and thus helps the materials manager to take timely action. It also helps to determine how the resources can be transferred from one job to another, thus providing alternative courses of action to accomplish the desired goal.

PERT involves the construction of a network of activities connecting the necessary events of a system in a sequence which culminates in the completion of the project. The network is a flow plan consisting of the activities and events that must be completed to reach the goal of the programme together with their interdepedencies and interrelationships. Thus, it shows orderly step-by-step series of activities carried out in a logical sequence to reach the goal. A very simple network indicating the decision on make or buy is given in Exhibit 33.1.

Events and Activity

An event in the PERT network is described as a clear step or milestone or control point or specific definable achievement in the plan. It dos not consume either time or resource. The event describes the beginning or the completion of one or more

EXHIBIT 33.1
A PURCHASING APPLICATION OF PERT MAKE/BUY DECISION

1	2 *Activities*		3 *Preceding event*	4 *Succeeding event*	5 *Activity times (Weeks)*	6	7	8	9 *Event times (Weeks)*	10	11 *Slack (Weeks)*
			i	j	t_o	t_m	t_p	t_e	T_E	T_L	
A.	Obtain detailed specification and prepare Engineering Drawings		1	2	2	7.5	16	8	8	8	—
B.	Preparing purchasing specifications and send enquiries for sub-assembly	one	2	4	2	4	12	5	13	13	—
C.	" " "	two	2	5	2	4	6	4	12	12	—
D.	Compute manufacturing costs for sub-assembly	one	2	3	1	3.5	9	4	12	14	2
E.	" " "	two	2	6	1.5	2.5	6.5	3	11	14	3
F.	Get quotation for sub-assembly	one	4	7	4	5.5	10	6	19	21	2
G.	" " "	two	5	8	4	8	12	8	20	20	—
H.	Review and approval of costs of sub-assembly	one	3	7	1	5	9	5	19	21	2
I.	" " "	two	6	8	3	6	9	6	20	20	—
J.	Evaluation of quotations for sub-assembly	one	7	9	1.5	3	4.5	3	22	24	2
K.	" " "	two	8	9	1	4	7	4	24	24	—
L.	Make or Buy decision		9	10	1	2	3	2	26	26	—

t_o = optimistic time for an event.
t_m = most likely time for an event.
t_p = pessimistic time for an event.
t_e = expected time of an event.

T_E = Earliest completion time of an event.
T_L = Latest completion time of an event.

activities; hence it should be tangible, meaningful, specific, and satisfactory to the person performing the events. The events are represented by circles, squares, rectangles in the network diagram. The events form the foundation for describing the activities included in between them.

An activity is work actually in process—resources being consumed or facilities being used with man-hours being spent. It is a task, job or assignment characterised by persons using resources for some period of time to achieve the stated objectives. An activity is represented by an arrow or line drawn between two events in the network diagram. Activities are the channels of flow of the network and the manager can control this flow of material, investments, expenses, human efforts, time spent, etc., utilised to achieve the final project goal.

Purchasing Application of PERT

Let us now consider an example in the purchasing field, namely, make or buy decision for an assembly consisting of 2 sub-assemblies. The final job of make/buy the assembly is broken into major components shown in column 2 of Exhibit 33.1. These are then drawn in the form of a network as in Exhibit 33.2. PERT is also useful in analysing Lead Time, expediting orders, etc. In this network given in Exhibit 33.1, the process of preparing an engineering drawing is an activity whereas the completion of the drawing marks an event. In the network, Exhibit 33.2—the events are numbered in a serial order and each activity is defined by the events preceding and succeeding it. For example, the activity—A—or the process of preparing the drawing is numbered as activity (1), (2). Here it is noted that event 1 signifies the approval of the policy and event 2 denotes the appearance of the completion of the drawing. It is also seen that the length of the line does not represent duration of the activity except in a time scale network. Any activity succeeding an event cannot be started until this event has occurred. For instance, for sub-assembly one cannot, in this network, start until the drawings are complete. Similarly, the event succeeding an activity cannot occur until all activities leading to it are accomplished. Looping, i.e., an event having an activity leading back to a preceding event, is not permitted in the network diagram. An event cannot occur twice or supersede itself. Further, it is noted that all the activity paths must be completed by the appropriate events. A full description of each activity should be written above the arrow linking the events. All the constraints, interdependencies should be clearly shown in the diagram.

Dummy activity is defined as an activity which does not consume time or resource but it is a useful and necessary constraint, completion of which depends upon further activities. The numbering of the events could be forward or backward depending upon whether it begins from the starting point or from the goal. Exhibit 33.2 adopts a forward numbering method.

Network Analysis

Since PERT is concerned basically with control of time dimension, the activity time or the duration associated with each activity is estimated as follows: Three time estimates—optimistic t_o, most likely t_m, and pessimistic t_p, are made available

EXHIBIT 33.2
MAKE OR BUY THROUGH PERT

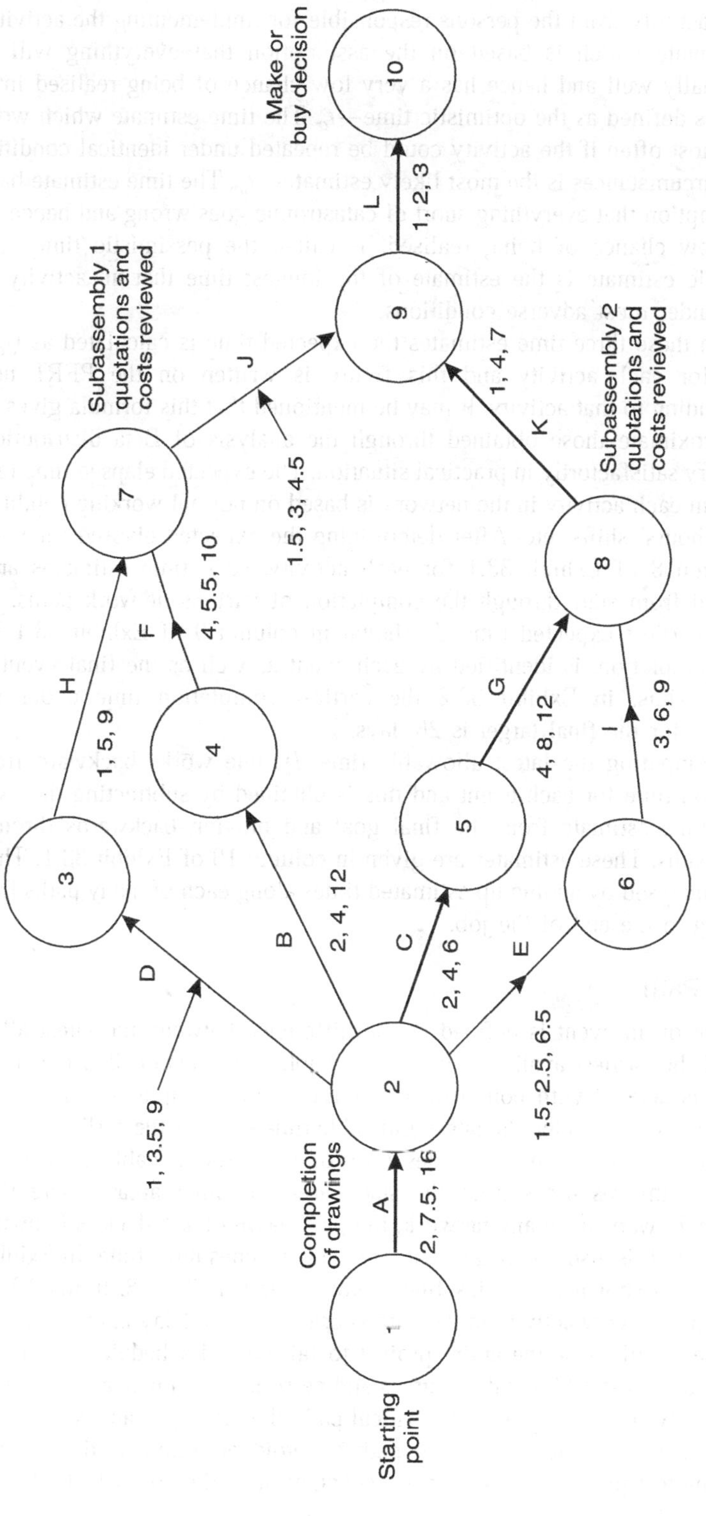

for each activity from the persons responsible for implementing the activity. The time estimate which is based on the assumption that everything will go on exceptionally well and hence has a very low chance of being realised in actual practice is defined as the optimistic time—t_o. The time estimate which would be correct most often if the activity could be repeated under identical conditions in normal circumstances is the most likely estimate—t_m. The time estimate based on the assumption that everything short of catastrophe goes wrong and hence having a very low chance of being realised, is called the pessimistic time—t_p. The pessimistic estimate is the estimate of the longest time that an activity would require under most adverse conditions.

From these three time estimates the expected time is calculated as $(t_o + 4t_m + t_p)/6$ for each activity and this figure is written on the PERT network corresponding to that activity. It may be mentioned that this formula gives results that approximate those obtained through the analysis of Beta distribution and works very satisfactorily in practical situation. The expected elapsed time required to perform each activity in the network is based on normal working conditions of working hours, shifts, etc. After determining the expected elapsed time—please see column 8 of Exhibit 33.1 for each activity, such time estimates are then cumulated from start through the completion of various network paths. In this way the earliest expected time T_E shown in column 9 of Exhibit 33.1 for the project completion, is identified by each event as well as the final event in the network. Thus, in Exhibit 33.2 the earliest completion time under normal conditions for the final target is 26 days.

In computing the latest allowable time T_L, one works backward from the completion time for each event and this is obtained by subtracting the expected elapsed time estimate from the final goal and moving backwards through the various paths. These estimates are given in column 10 of Exhibit 33.1. Thus, the chart is analysed by adding up estimated times along each of many paths from the beginning to the end of the job.

Critical Path

The slack of an event is defined as the difference between the latest allowable time and the earliest available time to do a job and is also called the float. The slack is associated with non-critical activities—for example—event 7. It is the number of days by which the latest allowable time exceeds the earliest completion time for the event. Hence, it provides a sort of elbow room enabling the executives to transfer the resources from the slack areas to other areas where they are immediately wanted. In any network, the path having least slack is known as the critical path. It is also the longest path which consumes most time. In Exhibit 33.2 this has been shown by a thick line linking events 1, 2, 5, 8, 9 and 10. As the name implies, every activity in the path is critical, for a delay in any of the events in this path will cause the entire project to fall behind schedule and thus defeat the purpose unless additional resources and personnel are made available to offset the delay. By concentration on the critical path, the manager can transfer facilities from the slack path to the critical path to avoid or minimise delays and thus ensure the test project time to remain constant; also the cost will be kept at the minimum. Here zero slack indicates that the project will be completed on schedule, positive slack indicates the project will be finished ahead of the completion date and the negative slack denotes that the event is likely to slip behind the scheduled date.

Thus, the network analysis consists of: (a) obtaining a work breakdown structure to a set of individual jobs which are arranged into a logical network, (b) estimation of job duration, and (c) scheduling to discover which jobs control the completion of the project and reallocation of resources to improve the schedule.

Computer

Some complex projects involve a large number of events and activities and hence are too tedious for manual calculations. For such large projects, the use of a computer may be advantageous. The computer could convert the days or weeks into calendar dates and also help to calculate the expected time, completion time, latest allowable time, activity slack, etc. It must be mentioned that the lack of a computer should not deter one from using the network.

Master Network

Each network is drawn for a specific level of management and reflects the degree of detail and interest to that particular level. For instance, the example discussed shows only one aspect namely make/buy. This has to be dovetailed into the entire project. The network for the entire project is known as Master Network. In most large-scale projects, it will be desirable to prepare a number of separate networks, describing the performances of different departments or individual portions of the project. The end events of such different networks are integrated and shown in a master network which gives the total project view. Thus, it is noted that the events in the master network are the end events of subsidiary networks, through which the top level policy and objectives can be clearly understood and enforced throughout the organisation. Events common to more than one network are known as interface events and such interface events must be identified before the networks are completed. It is also found advantageous to maintain a special network for the entire project which shows only the interface events.

Manpower Planning

From the earliest start and the latest allowable time for each event and also the estimated manpower needs for each activity, it is possible to have a balanced manpower employment throughout the duration of a project. The manpower required for each category during each week or month is calculated and plotted in the form of a bar chart. This bar chart indicates the manpower on the vertical axis and the calendar weeks or months on the horizontal axis. Then it is possible to identify the peaks and valleys in employment. With a knowledge of the activities on the critical path as well as the latest allowable time for the different events, it is possible to adjust the starting time of the different activities. By repeating such adjustments in alternative steps, it is possible to reach a balanced level of employment for each category. The above procedure helps to have a constant workforce throughout the construction phase of the project instead of adopting a policy of 'hiring and firing'.

Time-Cost Relationships

The time-cost trade off problem is directed to the cost of determining a schedule of project activities which considers adequately the indirect as well as direct costs and attempts to minimise the sum. By considering the cost feature of critical path methods, an accurate relationship can be developed between direct project costs and the time required to complete the project.

If the management desires to complete the job—(for instance, in our example on make/buy in less time than the 26 weeks)—earlier than the latest allowable time for the end event, then extra cost has to be incurred to expedite the job. For this purpose, the critical path has to be compressed and this process is called "Crash Programming". Here, the costs include direct, indirect and opportunity costs. The opportunity costs represent the penalty for not accomplishing the job at the earliest possible time. In the normal estimate, the primary emphasis is on cost with normal time being associated with minimum cost. The crash estimate involves the absolute minimum time required to do the job and the cost necessary to achieve that time with emphasis placed on time, after providing for overtime, extra shifts, etc. The crash activity time will be short but the programme cost will increase.

Analysis of the cost to expedite each activity in the critical path of a project provides a procedure to crash an activity which offers minimum increase in cost per week. Thus, a new schedule with more cost but less time can be obtained. At project level the activity with the lowest possible slopes should be crashed indicating that time should be purchased as cheaply as possible on the critical path. Thus, to crash the schedule, the time and cost estimates of critical path are reviewed and the resources from non-critical to critical path wherever possible are transferred to expedite the final event. For expediting the project further, the activity with lowest slope on the critical path is crashed.

PERT is also extremely useful for pre-crisis planning and buying when *force majeure* clauses are operating, i.e., to plan in advance and allocate responsibilities to the different executives in case of fire, strike, earthquake, major accidents, floods, etc., so that each executive knows what precisely his responsibilities are and what he has to do in such situations without having to await instructions.

In making the network plan valid, the manager himself must be involved in, and associated with, the preparation of the network. A network is never really finished until the project is finished. It is a dynamic living plan which needs to be revised as and when new difficulties are discovered. By anticipating the trouble spots likely to appear, application of PERT results in improved utilisation of resources and hence aids better managerial decisions. Therefore, the use of networks needs clear definition of objectives, thus forcing proper communication and enabling management by exception to be adopted.

34

Computers in Materials Management

ELECTRONIC COMPUTER

When the first electronic computer was installed in India in the early 1950's, it was generally believed that this device would replace most of the work of the clerical staff and management personnel. However, today the scene has changed considerably. Today, computers have penetrated into practically every aspect of life, particularly in management and information technology areas. Computers are being used by a large number of organisations. There are about 15 different brands of computers in use. Personal Computers (PCs) are extensively being used for various operations. Computers are employed both by private and public sectors, educational institutions, research organisations and government departments.

The maximum area of application is in payroll for processing the large mass of data in quick time with great accuracy, and financial accounting. Materials management is an area in which the computers can be put to beneficial use. Out of a possible 360 meter hours of computer in a month, only 25 hours are utilised for materials management. Since computerisation of inventory control will provide better control and profitability, there is rich potential for computerisation in materials management field.

INTEGRATED COMPUTER SYSTEM FOR MATERIALS MANAGEMENT

We can sub-divide the materials management function into Forecasting Prices, Materials Planning, Purchasing and Inventory Management.

These activities are inter-related and they are diagrammatically represented (see Exhibit 34.1). All these activities can be easily computerised.

MATERIALS PLANNING

This is the first activity in materials management. The important input here is the sales forecast. The sales forecast is converted into finished gods forecast, and the bill of materials is used for exploding the requirements of finished goods

EXHIBIT 34.1

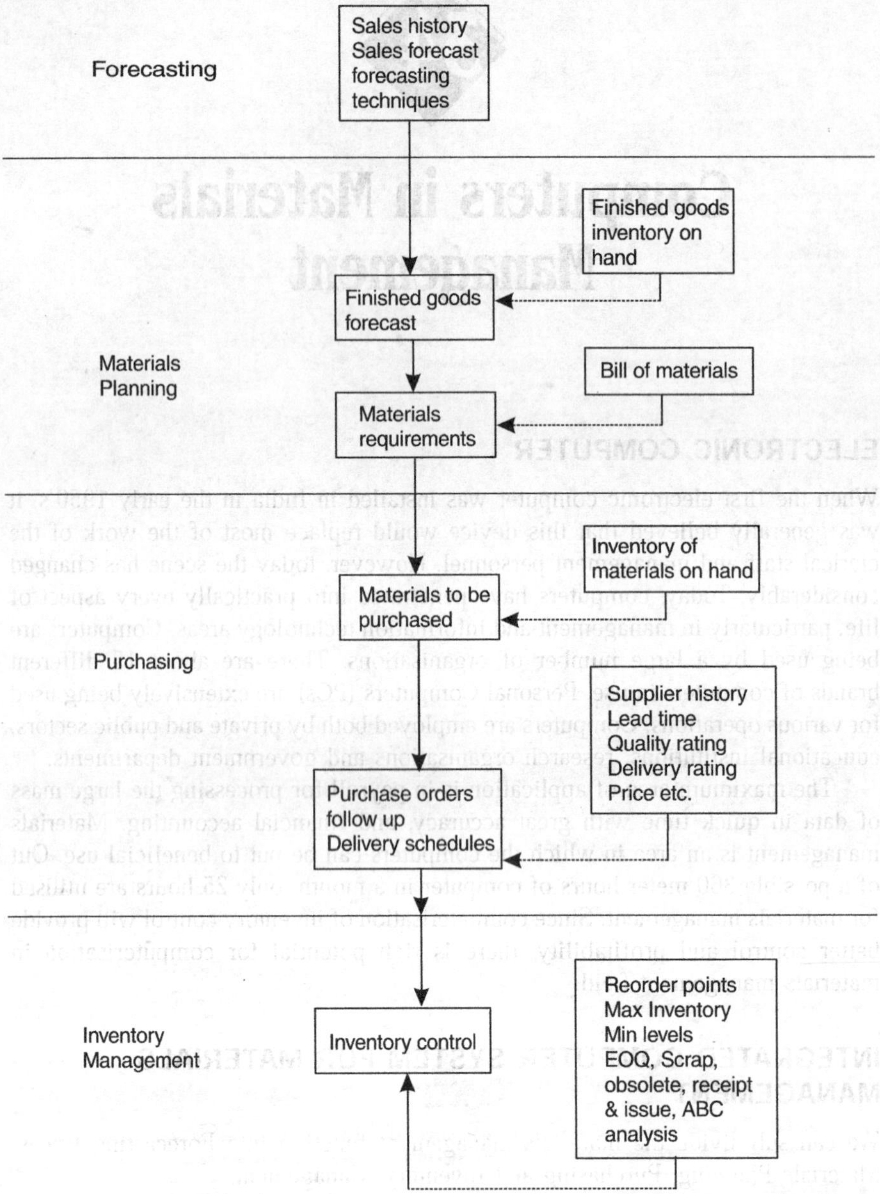

into the requirements of materials. The bill of materials is a document containing a detailed list of all raw materials, components and sub-assemblies needed for manufacturing a particular final product. It will be very exhaustive and will carry part numbers, store locations, whether they are to be purchased or manufactured, and so on. This can be easily explained by the following example (see Exhibit 34.2).

EXHIBIT 34.2

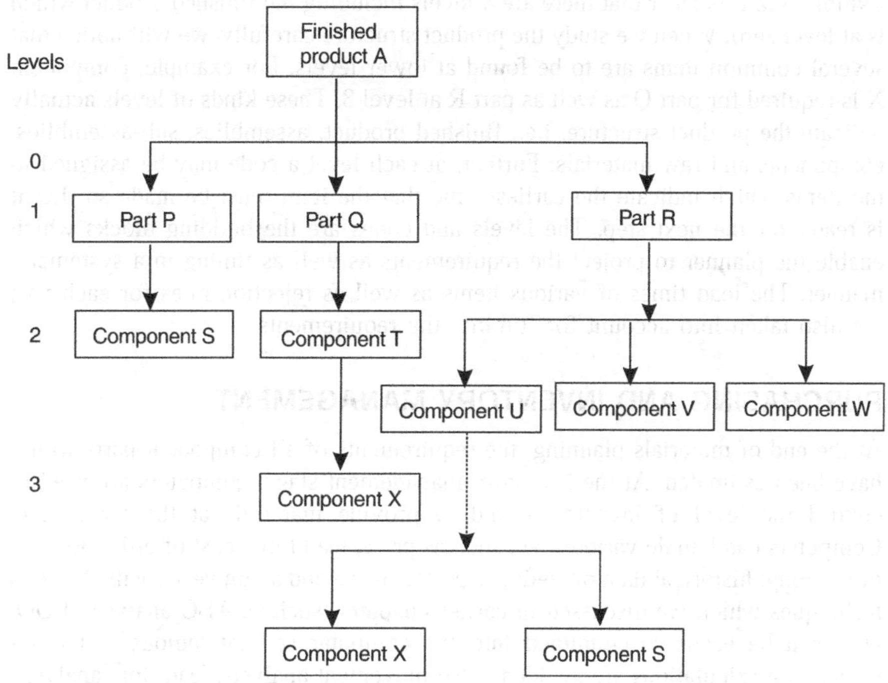

Bill of materials for A

		Requirements
Part	P	1
Part	Q	1
Part	R	1
Component	S	2
Component	U	1
Component	V	1
Component	W	1
Component	T	1
Component	X	2

When the forecast of finished product A for a month is 1,000 units, the bill of materials for A can be used to arrive at the forecast of the requirements of various parts and components which go into the manufacture of A. Bill of materials data are sometimes kept in punched cards. As the bill of materials are formed from the engineering drawings, any changes can be effected by replacing invalid cards with new cards. However, when the number of parts that go into each finished product becomes large (say 10,000 items or more as could be the case with Automobiles), then the punched cards could become unwieldy. Hence, many computer systems operate by storing the bill of materials internally in the machine.

Also, in computerised materials planning it is possible to send the exploded requirements over the future. This is done by assigning levels which is known as product structuring. Programmes are available to assign a level. Thus from Exhibit 34.2 it is clear that there are 4 levels including the finished product which is at level zero. When we study the product structure carefully, we will notice that several common items are to be found at lower levels. For example, component X is required for part Q as well as part R at level 3. These kinds of levels actually indicate the product structure, i.e., finished product, assemblies, sub-assemblies, components and raw materials. Further, at each level a code may be assigned to the items which indicate the earliest time that the item must be made so that it is ready for the next step. The levels and codes are the building blocks which enable the planner to project the requirements as well as timing in a systematic manner. The lead times of various items as well as rejection rates for each part are also taken into account for 'timing' the requirements.

PURCHASING AND INVENTORY MANAGEMENT

At the end of materials planning, the requirements of all component parts would have been estimated. At the inventory management stage, computers are used to control the level of inventories and to provide materials at the right time. Computers can handle various data such as price, lead time, cost of ordering, cost of carrying, historical data on delivery performance and so on very easily. Various techniques which we discussed in earlier chapters (such as ABC analysis, EOQ, etc.) can be easily programmed into the computer so that tedious and time-consuming calculations are avoided. Also movement analysis, lead time analysis, vendor rating, etc., can be computerised in a short time which permits the management to carefully evaluate and take scientific decisions so as to control the inventory levels. Factors, such as reserve stock, safety stock and reorder points, require statistical and mathematical analysis; when thousands of items have to be procured, it will be impossible to control them manually. Programmes are available for performing ABC analysis, EOQ calculations, reorder point computations and delivery schedule printing. This will come as a part of the "software package" in most of the cases. A-items are normally controlled manually with crucial and timely information support from computers. Thus, computers may print out the stock status of A-items, orders pending execution, expected consumption, etc., so that follow-up will be done on a selective basis to keep inventories at minimum. B and C items are usually large in number and here computers may be used to order EOQ and reorder point control can be used. Requisition analysis, processing of enquiries and quotation analysis can be done very effectively using computers. Receipt and issue documents can be easily computerized. This keeps in product costing, materials accounting and maintaining perpetual inventory records. This results in better physical verification.

With their increasing variety and the heavy demand made on them, materials are today assuming a very important role in the operational efficiency of an organisation. It is imperative to conclude that computers could be utilised to guide all the activities involved in getting the materials to the factory. This integrated computer system would involve the active participation of line personnel and would be more useful for operational information—management reports could however be obtained as a by-product.

The present use of computers in materials management is only to make computations easier and faster. The integrated computer system should be utilised for the sake of an integrated information system which would result in the smooth operation of the materials functions. The main objective of such an integrated system should be accuracy in reporting demand and inventory. The integrated system also ensures a fast reaction to environmental changes—changes in either supply or price of materials or changes in customer taste and measurement of functions such as inventory, purchasing and vendors.

The normal flow of information regarding materials in an organisation is depicted below with detailed explanation of list of activities.

A computer can easily maintain this information flow with basic data, coming from production planning, design and maintenance departments and environmental information coming from purchase and sales departments. It can give back the formation of inventory records to the production planning department, can send purchase requisitions to the purchase department, remind purchase department for follow-up and place direct purchase orders for standard repetitive items. It can update the past record file with revised consumption figures for items, and can check the actual consumption against the bill of materials and debit material cost incurred to user departments. On the basis of record and basic data, the computer can formulate inventory levels.

The distinct advantages of an integrated computer system are: that it involves the line personnel and provides timely and quick information on the materials position. It is also used for operational convenience and not for mere calculations. Studies conducted in India and abroad indicate that computers have been fruitfully used in organisations where the primary objective of the computer system is operations management. The use of computers for only computations is probably the reason why businessmen are not sure of their worth.

Integrated Management Information Systems (Exhibit 34.3) using computers are applicable in the following areas. The list is only illustrative and not exhaustive.

1. Long-term production schedule.
2. Short-term production schedule.
3. Materials manual.
4. Requirement of non-stock items from user departments.
5. Information regarding lead time, supply position (shortages), price trends, anticipated price changes, etc.
6. Production schedule handed over to production department.
7. Materials requisition from the production to stores.
8. Materials supplied from stores to production.
9. To and fro information between stores and inspection.
10. Information regarding receipts from the stores.
11. Date rcgarding issues from stores.
12. Due dates of supply from purchase department.
13. Information to purchase department for follow up of supplies.
14. Purchase requisition to purchase department.
15. Purchase order.
16. Materials from suppliers to stores.
17. Previous year's consumption data.

EXHIBIT 34.3

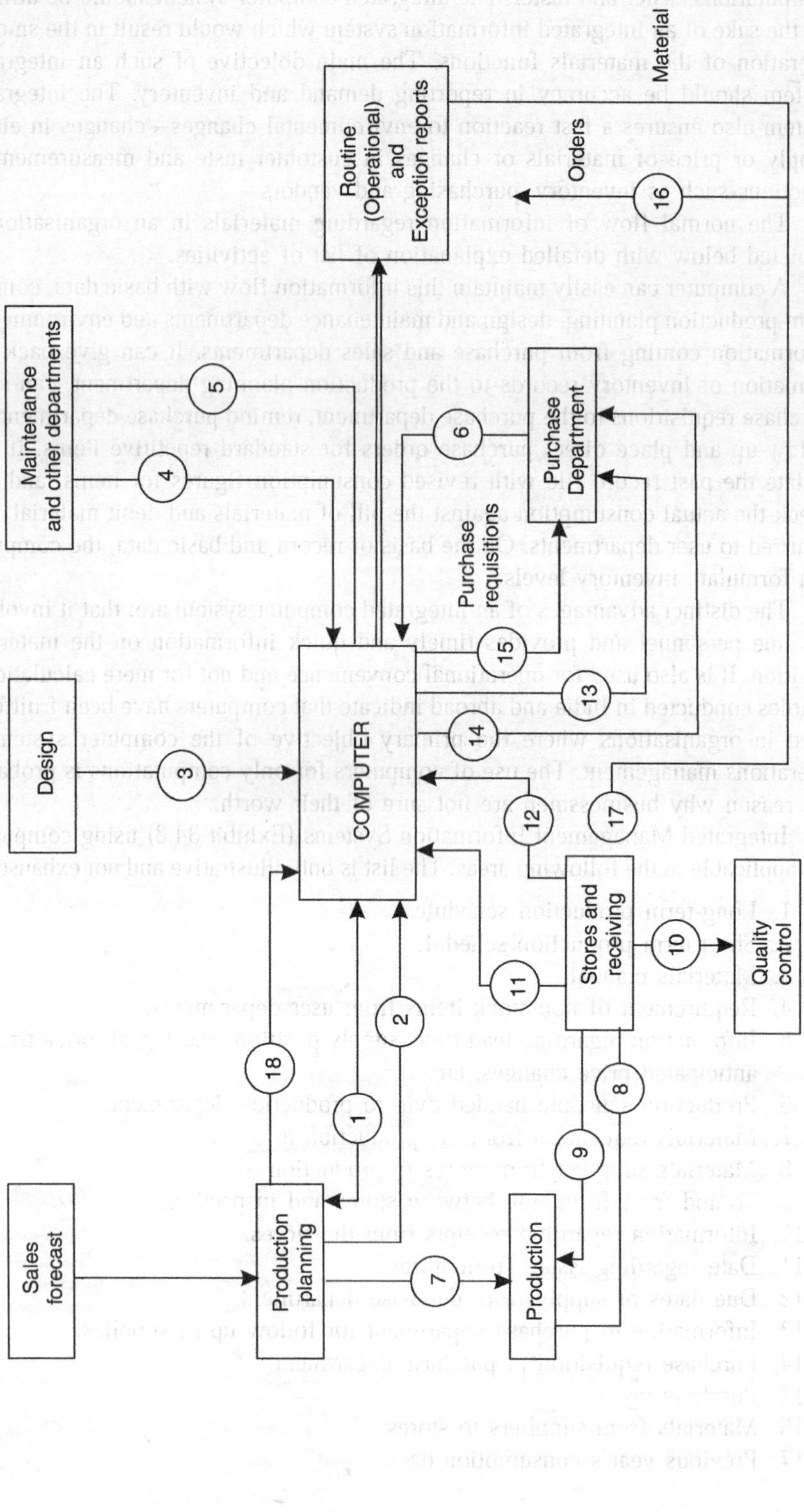
Sales forecast
Production planning
Production
Design
Maintenance and other departments
COMPUTER
Routine (Operational) and Exception reports
Purchase requisitions
Purchase Department
Orders
Material
Stores and receiving
Quality control
1
2
3
4
5
6
7
8
9
10
11
12
13
14
15
16
17
18

OPERATIONAL AND EXCEPTIONAL REPORTS

The following operational reports emanate from the computer. Here again the list is not exhaustive.

1. Bill of materials.
2. Price forecasts.
3. Purchase budgets.
4. ABC analysis.
5. Inventory levels.
6. Items below safety stock levels.
7. Movement analysis.
8. Materials accounting and summary of issues.
9. Stores accounting.
10. Stock verification.
11. Vendor rating.
12. Sectorwise and material categorywise purchase list.
13. Outstanding payments.

It is possible to get the following exceptional reports for managerial purposes.

1. Items for which consumption is more than the norms.
2. Items for which there is a supply shortage and anticipated price increase.
3. Items out of stock.
4. Items with stock level above maximum norms.
5. Obsolete items.

CONCLUSION

In computer systems, the basic law of the use of computers is explained by acronyms such as GIGO (Garbage in : Garbage out) and TITO (Trash in : Trash out). Therefore, a disciplined data base is absolutely essential for the successful operation of the computer systems. Many developments, such as time-sharing concept, process control, on line systems concept, teleprocessing, etc., have been included in the field of computers. In India, third generation computers with sophisticated integrated circuits are now commecially sold. Therefore, these factors have to be carefully evaluated before deciding on computerisation; the merits involved in renting the facilities rather than buying the system outright also deserve careful attention. A thorough cost-benefit analysis must be made before arriving at a decision.

35

Evaluation of Materials Management

The focus of materials management has been presented in the earlier chapters as a result-oriented approach. In the area of inventory control and cost reduction, the benefits are measurable and immense. However, there is a need for an objective evaluation of the performance of materials management. The top management should spot and recognise the competence and ruthlessly weed out incompetence. The rewards and punishment should be commensurate with the results achieved, so as to infuse motivation and fix responsibilities. Such responsibilities will be both a challenge and opportunity for the materials manager as the field for savings and cost reduction is indeed high.

ORGANISATION

The evaluation of materials management can either be done by external agencies or internally. Within the organisation itself, the top management, usually at the end of the financial year and periodically during the year, evaluates the performance on the basis of inventory holding and obsolete items. The user-departments evaluate the materials function in terms of the number and the duration of the stock-out. In a few Indian organisations, committees frequently review the performance of materials management against the objectives set for the department.

It is known that suppliers who are evaluated by the materials management department, in turn, evaluate the materials manager with regard to quality consciousness, adherence to payment schedule and his importance as well as decision-making capabilities in the hierarchy. The other external agencies interested in the performance of a materials manager are: banks with regard to the credit-worthiness, and professional associations and national association of materials management with respect to the ethical practices. Occasionally, external consultants evaluate the materials management systems to suggest improvements.

DIFFICULTIES

Before proceeding to discuss the process of evaluation, it is desirable to understand the background in which the materials manager operates. While all the managerial functions like production, marketing, etc., as well as a large number of other professions in the world are evaluated on the basis of their achievements, materials management is one of the few professions which is judged by failures or stock-outs. It should also be emphasised that the integrated concept of materials management is hardly a decade old and hence there could be misjudgments on the part of the materials manager. In a controlled economy like India's, it is increasingly difficult to procure materials as per well-known principles. The lead time is invariably high because of administrative controls and procedures. What is worse in some firms, the materials manager is held responsible for the total inventory, including raw material, work-in-progress, finished goods and obsolete items, even though he can control only the raw materials. In view of all this, goals are difficult to set, hard to achieve and hence impractical to measure, as the ideal conditions do not exist in practice. The concept of MBO or management by objectives discussed in Chapter 2, can also be used for goal setting and evaluation purposes.

PROCESS AND CRITERIA

The process of evaluation can be visualised on various measures, either on temporal comparison basis, i.e., performance over financial or calendar years—or between similar companies—inter-firm comparisons—over the same year. The criteria of evaluation, obviously, depend on the corporate objectives, company culture, functions of materials management, evaluation process of other departments, etc., and hence could vary, and do vary, from company to company. An illustrative list of criteria in the form of ratios and questions is given at the end of the chapter. The process in all the cases is to formulate norms for each of the characteristics and compare the actual performances with the norms. It should be emphasised, in this context, that it will not be possible to set norms for subjective characteristics.

The subjective characteristics are source development, training and man-power development as well as morale in materials department, exception reporting, good buyer-seller relations, satisfaction to the user, materials managers' knowledge of processes and products, attitude to modern techniques of materials management and development of manuals and procedures. The objective criteria include the frequency and intensity of stockouts, ratios involving inventory such as sales to inventory and value of obsolete items, consumption to inventory carrying cost, transport cost, costs of acquisition, use of codification and value analysis, variety reduction and standardisation.

REPORTING

Evaluation is also done by the top management on the basis of the periodic reports received from the materials management department. Communication is the life-blood of business management. Collection and dissemination of data are very essential in reporting. Mere facts can reflect what has happened whereas reports would give the summary of the working of the materials department. Monthly reports consisting of purchases, inventory, stockouts and cost reduction achieved through materials department are common in industry. In order to take timely corrective measures, the monthly reports should be submitted during the first week of the following month. Quarterly and annual summaries, with specific emphasis on long-term aspects of materials management, such as source development and value analysis implementation, are also prepared by the materials department. The important aspects in reporting are: (a) clarity, (b) brevity, and (c) periodicity (related to frequency.)

Given below are various ratios which can be used to evaluate the performance of different functions of materials management.

PURCHASING

Materials cost index

1. $\text{Material cost index} = \dfrac{\text{Material cost for a product}}{\text{Production cost of the product}}$

2. $\text{Material cost/sales index} = \dfrac{\text{Material cost for each product}}{\text{Sales value}}$

3. $\text{Indigenous content index} = \dfrac{\text{Cost of indigenous material}}{\text{Total material cost}}$

4. $\text{Material variances index} = \dfrac{\text{Value of material actually used}}{\text{Value of standard material quantity}}$

Vendor rating indexes

5. $\text{Vendor rating index (Quality)} = \dfrac{\text{No. of lots rejected}}{\text{No. of lots received}}$

6. $\text{Vendor rating index (Delivery)} = \dfrac{\text{Delivery on schedule}}{\text{Total no. of deliveries}}$

7. $\text{Vendor rating index (Price)} = \dfrac{\text{Lowest price bid}}{\text{Price bid by vendor}}$

8. Vendor rating index (Overall) = VRI (quality) × A + VRI (Delivery) × B + VRI (Price) × C, where A, B and C are the weights given to the three vendor rating indexes by the materials manager

Material planning and forecasting indexes

9. $\text{Forecast price index} = \dfrac{\text{Forecast price of material}}{\text{Present price of material}}$

10. $\text{Forecast accuracy index} = \dfrac{\text{Forecast price of material}}{\text{Actual price of material}}$

11. Cash discount index (Quantity) = $\dfrac{\text{Cash discount availed of due to quantity buying}}{\text{Total value of purchase}}$

12. Cash discount index (prompt payment) = $\dfrac{\text{Cash discount availed of by prompt payment}}{\text{Total value of purchase}}$

13. Cash discount index = $\dfrac{\text{Total cash discounts availed of}}{\text{total value of purchase}}$

14. Unutilised cash discount index = $\dfrac{\text{Total cash discounts availed of}}{\text{Total cash discounts available}}$

15. Planning efficiency index = $\dfrac{\text{Number of rush orders placed}}{\text{Total no. of orders}}$

16. Price variance index = $\dfrac{\text{Standard price} - \text{actual price}}{\text{Standard price}}$

17. Insurance cost index = $\dfrac{\text{Total cost of insurance}}{\text{Total value of insurance}}$

18. Insurance index = $\dfrac{\text{Value of uninsured items}}{\text{Value of insured items}}$

19. Lead time index = $\dfrac{\text{Average lead time this year}}{\text{Average lead time last year}}$

Efficiency indexes

20. Manpower strength index = $\dfrac{\text{Total staff in purchase}}{\text{Total company staff}}$

21. Manpower cost index = $\dfrac{\text{Wage bill of purchase department}}{\text{Total wage bill of company}}$

22. Order cost index = $\dfrac{\text{Total purchase department cost}}{\text{Total no. of orders placed}}$

23. Rush order cost index = $\dfrac{\text{Price paid for rush order material}}{\text{Price normally paid for this material}}$

24. Ordering efficiency index = $\dfrac{\text{No. of orders of small Rs value (e.g., less than Rs 100)}}{\text{Total no. of orders}}$

25. Clerical time per order = $\dfrac{\text{Total clerical hours available per month}}{\text{Total no. of orders per month}}$

26. Clerical efficiency index = $\frac{\text{No. of clerical errors in posting, orders, etc}}{\text{No. of clerical tasks performed}}$

27. Process time index = $\frac{\text{Total time taken to process indents}}{\text{Total no. of indents processed}}$

28. Purchase efficiency = $\frac{\text{Total purchase value}}{\text{Total expenses of purchasing department}}$

29. Communication cost index = $\frac{\text{Cost of telephones, telex, postage}}{\text{Value of purchases}}$

30. Vendor development index = $\frac{\text{Total no. of vendors}}{\text{Total no. of items}}$

31. Time utilisation index = $\frac{\text{Time spent on calls with vendors}}{\text{Total man-hours available}}$

32. Claims index = $\frac{\text{Total claims obtained}}{\text{Total amount claimed}}$

33. Negotiation efficiency index = $\frac{\text{Actual price paid by the organisation}}{\text{Market price}}$

34. Indigenous source development index = $\frac{\text{Value of imported items}}{\text{The value of materials consumed}}$

35. Process efficiency index = $\frac{\text{No. of unprocessed indents}}{\text{Total no. of indents received}}$

36. Indents confirmed as per requirement index = $\frac{\text{No. of deliveries confirmed as per indent}}{\text{Total no. of indents}}$

37. Sales/Purchase index = $\frac{\text{Total sales}}{\text{Total purchases}}$

Inventory and stores

Raw material, spare parts, finished goods inventory and work-in-progress indexes.

38. Week's inventory on hand ratio = $\frac{\text{Material in stock}}{\text{Weekly consumption}}$

39. Week's inventory on order = $\frac{\text{Material on order}}{\text{Weekly consumption}}$

40. Inventory turnover ratio (finished goods) = $\frac{\text{Annual sales}}{\text{Average inventory}}$

41. Inventory turnover ratio (Raw Material) $= \dfrac{\text{Annual consumption}}{\text{Average inventory}}$

42. Priority or importance (number) index $= \dfrac{\text{No. of A class items}}{\text{Total no. of items}}$

43. Priority or importance index (value) $= \dfrac{\text{Value of class inventory items}}{\text{Total inventory}}$

44. Out of stock index $= \dfrac{\text{No. of times out of stock}}{\text{Number of times requisitioned}}$

45. Production loss due to stock out $= \dfrac{\text{Hours of production schedule not met due to material non-availability}}{\text{Total scheduled production time}}$

46. Work-in-progress index $= \dfrac{\text{Average work-in-progress}}{\text{Total production value annually}}$

47. Spare parts index $= \dfrac{\text{Value of spare parts inventory}}{\text{Value of capital goods}}$

48. Spare parts turnover index $= \dfrac{\text{Spare parts consumed}}{\text{Spare parts inventory}}$

Stores indexes

49. Handling cost index $= \dfrac{\text{Total handling cost}}{\text{Total value of material received and issued}}$

50. Handling loss index $= \dfrac{\text{Total value of losses due to handling}}{\text{Total value mat. rec. and issued}}$

51. Stores efficiency index $= \dfrac{\text{No. of requisitions delivered on time}}{\text{Total no. of requisitions}}$

52. Index of count effectiveness for bulk items

$$= \frac{\text{No. of items by count}}{\text{No. of items by weight, volume, etc.}}$$

53. Index of confidence level for book inventories $= \dfrac{\text{No. of book inventories within Y\% of physical count}}{\text{No. of book inventories verified by physical count}}$

54. Storage loss index = $\dfrac{\text{Value of inventory lost due to deterioration, obsolescence, pilferage, etc.}}{\text{Average value of inventory}}$

55. Obsolescence index (number) = $\dfrac{\text{Number of non-moving items}}{\text{Total no. of items in inventory}}$

56. Obsolescence index (value) = $\dfrac{\text{Value of non-moving items in inventory}}{\text{Total inventory value}}$

57. Scrap disposal index = $\dfrac{\text{Value of scrap disposed}}{\text{Total value of scrap}}$

58. Scrap loss index = $\dfrac{\text{Value of scrap sold}}{\text{Value at which bought}}$

59. Space utilisation index = $\dfrac{\text{Area used for storage}}{\text{Total storage area available}}$

60. Quality control effectiveness index (Incoming Inspection) =

$$\dfrac{\text{Total value of items rejected on inspection}}{\text{Total value of defective material}}$$

Standardisation, codification and value analysis

61. Value analysis effectiveness index = $\dfrac{\text{Savings made by value analysis}}{\text{Total material costs}}$

62. Standardisation effectiveness index = $\dfrac{\text{No. of items after standardisation}}{\text{Total no. of items before standardisation}}$

63. Codification index = $\dfrac{\text{No. of items codified}}{\text{Total no. of items}}$

64. Standardisation index = $\dfrac{\text{No.of items similarly functioning but with different specifications}}{\text{Total no. of items}}$

Transportation and Distribution Indexes

65. Freight index = $\dfrac{\text{Lowest rate bid for freight}}{\text{Rate paid}}$

66. Vehicle usage index = $\dfrac{\text{Actual trucking hours}}{\text{Total trucking hours available}}$

67. Vehicle capacity utilisation index $= \dfrac{\text{Total load shipped}}{\text{Capacity} \times \text{No. of trips}}$

68. Vehicle maintenance index $= \dfrac{\text{Total time spent on maintenance}}{\text{Total trucking hours available}}$

69. Capacity index $= \dfrac{\text{Total shipments by Company truck}}{\text{Total shipments}}$

70. Economic index for vehicles $= \dfrac{\text{Cost/ton by own shipment}}{\text{Cost/ton by freight carriers}}$

71. Carrier rating index $= \dfrac{\text{No. of shipments delivered on time}}{\text{Total no. of shipment}}$

72. Delivery schedule index $= \dfrac{\text{No. of orders delivered on time}}{\text{Total no. of deliveries}}$

73. Cost/Mile/Ton $= \dfrac{\text{Total cost of transport department}}{(\text{Weight shipped} \times \text{miles shipped})}$

74. Distribution loss index $= \dfrac{\text{Distribution cost}}{\text{Sales value of items distributed}}$

75. Goods in transit inventories index $= \dfrac{\text{Average value of F.G. in transit}}{\text{Average value of F.G. inventories}}$

76. Goods in transit index $= \dfrac{\text{Average value of F.G. in transit}}{\text{Total sales value of shipments over a period}}$

The evaluation process is also aided by investigating the following questions and policy decisions.

77. Is the department adequately staffed?
78. Have authority and responsibility been properly delegated?
79. Are the employees competent and trained to do their jobs?
80. Are the relations with the vendors satisfactory?
81. Does the department keep track of economic movements such as GNP, per capita income, etc.?
82. Does the department keep track of the labour situations and financial position, etc., in supplier organisations?
83. Is the system for follow-up and chasing effective?
84. Does the inventory control department have a system which automatically indicates the materials out of stock or below accepted minimum stock levels?

85. Has any O & M study been undertaken to reduce paper work and streamline the system to fulfil present and future needs?
86. Does the department keep track of research studies which have a direct bearing on the materials needed by your product?
87. Do the executives spend proportionally more time on A class items than on B & C class items?
88. Is your filing system efficient or does it take a long time to trace papers, etc.?
89. Is your information control system effective?
90. Does it pinpoint areas which require investigation?
91. Is a constant effort made to develop suppliers and provide technical help, etc., when required?
92. Is a constant effort made to keep informed of new products and processes?
93. Does the company have an integrated approach towards Materials Management?
94. Is a constant effort made to study the market prices of sub-assemblies, etc., which are presently made by the organisation in order to achieve cost reduction through make/buy decisions?
95. Have your inventory policies, viz., safety stock, order quantity, etc., been laid down on experience and intuition or through careful analysis of demands and supply patterns?
96. Have you achieved goal congruence between company and employee objectives?
97. Are your material requirements for new products planned sufficiently in advance?
98. Do you have a budgetary control system with periodic review?
99. Do you have strict policies laid down on use of telephones, telegrams, etc., to control communication costs?
100. Are there strict policies on acceptance of gifts, etc., by buyers in order to avoid charges of graft?
101. Have you developed drawings, etc., or do you buy only by samples?

Bibliography

1. BOOKS

A. Materials Management—General

American Management Association, *Managing the Materials Function—Tools, techniques, and company practices,* N.Y., the Association, 1959, 131 pp.

Ammer, D.S. *Materials Management,* Rev. ed., Homewood, Illinois, Irwin, 1968, xi, 607 pp.

Baily, P. and Farmer, D. *Managing Materials in Industry,* London, Gower Press, 1972, xv, 341 pp.

Ballot, R.B. *Materials Management—Result approach,* N.Y., American Management Association, 1971, x, 309 pp.

Bocchino, W.A. *Materials Management—Tools and techniques,* London, Industrial and Commercial Techniques Ltd., 1971, ii, 109 pp.

Brady, G.S. *Materials Handbook—An encyclopedia for purchasing managers, engineers, executives, and foremen,* 10 ed., N.Y., McGraw-Hill, 1971, viii, 1045 pp.

Compton, H.K. *Supplies and Materials Management,* London, Business Books, 1968, x, 439 pp.

Datta, A.K. *Rationalising Materials Management,* Calcutta, World Press, 1970, xxiv, 170 pp.

Gokarn, P.R. *Notes on Materials Management,* Hyderabad, Administrative Staff College of India, 1966, 58 pp.

__________, *Essentials of Materials Management,* Bombay, Somaiya, 1970, x, 160 pp.

India. Bureau of Public Enterprise. *Guidelines for Materials Management in Public Enterprises,* New Delhi, Manager of Publications, 1971, 11, 143 pp.

India. Lok Sabha, 3. Committee on Public Undertakings, *Report—Materials management in public undertakings,* Delhi, Lok Sabha Secretariat, 1967, vi, 90 pp. (Its Report No. 40).

India. Lok Sabha, 4. Committee on Public Undertakings, *Report—Action taken on the 40th report of the Committee on public undertakings* (3rd Lok Sabha). Delhi, Lok Sabha Secretariat, 1969, vii, 80 pp. (Its Report No. 38).

Lee, L. and Dobler, D.W. *Purchasing and Materials Management—Text and cases,* 2nd ed., N.Y., McGraw-Hill, 1971, ix, 850 pp.

McElhiney, P.T. and Cook, R.I. (Eds), *Logistics of Materials Management—Readings in modern purchasing,* Boston, Houghton Mifflin Co., 1969, xiii, 385 pp.

Mathew, P.D. *Handbook of Materials Management*, Calcutta, McHarty International, 1968, xx, 291 pp.

Menon, P.G., *Materials Management and OR in India*, New Delhi, M.M.J. Publications, 1968, 347 pp.

Mitchell, H. *Codification in Manufacturing and Materials Management*, Boston, Massachusetts, Cahners Books, 1973, 797 pp.

Nair, N. K. *Materials Management*, Bombay, Asia Publishing House, 1967, x, 347 pp.

Palit, A.R. *Outlines of Materials Management*, Bombay, Academic Press, 1970, 200 pp.

Peckam, H.H. *Effective Materials Management*, Englewood Cliffs, N.J., Prentice-Hall, 1972, 238 pp.

Plossl, G.W. and Wight, O.W. *Material Requirements Planning by Computer*, Washington, D.C., American Production and Inventory Control Society, 1971.

Sims, E.R. *Planning and Managing Material Flow*, Boston, Massachusetts, Industrial Education Institute, 1968, 440 pp.

Stelzer, W.R. *Materials Management*, Englewood Cliffs, N.J., Prentice-Hall, 1970, 223 pp.

Vandemark, R.L. '*Managing Material Control*, Dallas, Vandemark Inc., 1970, 184 pp.

_____, *New Ideas in Materials Management*, Dallas, Vandemark Inc., 1967, 216 pp.

B. Purchasing

Aijian, G.W. *Purchasing Handbook—Standard reference book on purchasing policies, practices, contracts and forms*, N.Y., McGraw-Hill, 1958, Various pages.

American Management Association, *Purchasing Research—The concept and its value*, N.Y., the Association, 1962, (Its Bulletin, 17).

Anyon, G.H. *Managing an Integrated Purchasing Process*, N.Y. Holt, Rinehart and Winston, 1963, 218 pp.

Bagot, K. *Profit Potential and Purchasing Negotiation*, N.Y., American Management Association, 1968.

Baily, P.J.H. *Purchasing and Supply Management*, 2nd ed., London, Chapman & Hall, 1969, 308 pp.

_____ and Farmer, D. *Purchasing Principles and Techniques*, London, Pitman, 1968, xi, 321 pp.

_____ and Tavernier, G. *Design of Purchasing Systems and Records*, London, Gower Press, 1970, xiv, 150 pp.

Barlow, C.W. *Purchasing for the Newly Appointed Buyer*, N.Y., American Management Association, 1970, 108 pp.

Berry, H.A. *Purchasing Management*, Waterford, Conn., National Foremen's Institute, 1964, 176 pp.

Blitz, J.F. *Policies and Techniques for Progressive Purchasing*, London, Industrial & Commercial Techniques Ltd., 1971, Various pages.

Bolton, R.A. *Systems Contracting—A new purchasing technique*, N.Y., American Management Association, 1966, 125 pp.

Brand, G.T. *Industrial Buying Decision*, London, Associated Business Programmes Ltd., 1972, 134 pp.

Brand, G.T. *Purchasing Decision Processes,* London, Cassell, 1972.

Cantor, J. *Evaluating Purchasing Systems,* N.Y., American Management Association, 1970, 78 pp.

Colton, R. *Industrial Purchasing,* Columbus, Ohio, Charles Merril, 1962, 525 pp.

Combs, P.H. *Handbook of International Purchasing,* Boston, Massachusetts, Cahners Books, 1971, viii, 168 pp.

Compton, H.K. *Glossary of Purchasing and Supplies Management Terms,* London, Purchasing Officers Association, 1965, 160 pp.

Curley, T.J. *Purchasing and Supply Management,* Sydney, West Publishing Corp., 1968, 382 pp.

Dan, D.M., *Principles of Purchasing,* Calcutta, Biswas Publishing House, 1973, 16, 279 pp.

Dand, R. and Farmer, D.H. *Purchasing in the Construction Industry,* London, Gower Press, 1970, 198 pp.

DeRose, L.J. *Negotiated Purchasing—The key to more profitable buying,* Boston, Massachusetts, Materials Management Institute, 1962.

Dowst, S.R. *Basics for Buyers—A practical guide to better purchasing,* Boston, Massachusetts, Cahners, 1971, 215 pp.

England, W.B. *Modern Procurement Management,* Homewood, Illinois, R.D. Irwin, 1970.

__________, *Procurement—principles and cases,* 4th ed., Homewood, Illinois, Irwin, 1962, 947 pp.

__________, *Purchasing System,* Homewood, Illinois, Irwin, 1967, vii, 156 pp.

Fearon, H.E. *Purchasing Economics,* N.Y., American Management Association, 1968, 16 pp.

__________ and Hoagland, J.H. *Purchasing Research in American Industry,* N.Y., American Management Association, 1963, 80 pp.

General Services Administration, *Washington: Procurement handbook,* Washington, DC, the Administration, 1959, vii, 271 pp.

Great Britain, Department of Employment, *Training for Purchasing and Supply,* London, HMSO, 1972, 55 pp.

Gross, M. *Make or Buy,* Englewood Cliffs, N.J., Prentice-Hall, 1966, vii, 231 pp.

Haas, G.H., Krech, E.M., March, B. *Purchasing Department Organisation and Authority,* N.Y., American Management Association, 1960, 119 pp. (AMA Research Series, 45).

Harvard Business Review: *Procurement Series*—7 reprints from Harvard Business Review, 1950–62.

Hayes, F.A. and Renard, G.A. *Evaluating Purchasing Performance,* N.Y., American Management Association, 1964, 125 pp. (AMA Research Study, 66).

Hedges, H.G. *Procurement—The modern science of purchasing,* N.Y., Harper & Brothers, 1961, vii, 405 pp.

Heinritz, S.F. and Farrell, P.V. *Purchasing—Principles and Applications,* 5th ed., Englewood Cliffs, N.J., Prentice-Hall, 1971, xi, 460 pp.

Henrick, F.D. *Purchasing Management in the Smaller Company,* N.Y., American Management Association, 1971, 190 pp.

India. Stores Purchase Committee: *Report,* March 15, 1955. Delhi, Manager of Publications, 1959, iv, 239 pp.

Institute of Purchasing and Supply, London. *Seminar and Courses for Purchasing and Stores Personnel,* October 1968, 39 pp.

Japan Productivity Center. *Materials and Purchasing Control in Japan,* Tokyo, the Centre, 1960, 20 pp.

Jennings, G.W. *State Purchasing—The essentials of a modern service for modern government,* Lexington, Kentucky, the Council for State Governments, 1969, 118 pp.

Kay, F. *Purchasing—For industry and public undertakings,* London, Pitman, 1963, v, 229 pp.

Kollios, A.E. and Stempel, J. *Purchasing and EDP with a Case Study,* N.Y., American Management Association, 1966, 128 pp.

Leenders, M.R. *Improving Purchasing Effectiveness Through Supplier Development,* Boston, Division of Research, Graduate School of Business Administration, 1965, x, 152 pp.

Levitt, T. *Industrial Purchasing Behavior—A study of communication effects,* Boston, Massachusetts, Harvard University, 1965, vi, 184 pp.

McMillan, A.L. *Art of Purchasing,* N.Y., Exposition Press, 1959, 385 pp.

__________, *How to Teach Purchasing—Outline plans for study courses in purchase and supply,* N.Y., Exposition University, 1959, 37 pp.

Manente, M. *Functions of the Purchasing Manager,* N.Y., American Management Association, 1969, 30 pp. (AMA Management Bulletin, 132).

Mathews, L.M. *Increasing Purchasing Effectiveness,* London, Industrial and Commercial Techniques Ltd., 1972, 89 pp.

Meynaud, J. *Better Buying Through Consumer Information,* Paris, Organisation of European Economic Community, 1961, 136 pp.

Nicholson, J.W. and others: *Guide to Government Purchasing,* Lakewood Publications, 1965.

Pegram, R.M. *Purchasing Practices in the Smaller Company,* N.Y., Conference Board, 1972, 34 pp.

Pinkerton, R.L. *Curriculum for Purchasing,* Madison, Wisconsin, University of Wisconsin, Bureau of Research and Service, 1969.

Pooler, V.H. *Purchasing Man and His Job,* N.Y., American Management Association, 1964, 426 pp.

__________ and Johnson, R.G. *Fundamentals of Effective Purchasing,* N.Y., Argyle, 1967.

Purchasing Officers Association. *Purchasing Problems—A symposium of case studies and exercises,* by Peter, J.H. Baily and D.H. Farmer. London, the Association, 1967, 132 pp.

Reck, D., *Government Purchasing and Competition,* University of California Press, 1954, 215 pp. (California University Bureau of Business & Economic Research Publications).

Smith, A.E. *New Techniques for Creative Purchasing—A handbook for purchasing people who want to upgrade their jobs from routine service to the management level of operations,* Chicago, Dartnell Corp., 1966, 160 pp.

Trux, W.R. *Data Processing for Purchasing and Stock Control,* London, McGraw-Hill, 1971, x, 368 pp.

United Nations Industrial Development Organisation: *Guide to Industrial Purchasing*, N.Y., The Organisation, 1972.

Webster, F.E. *Organisational buying behavior*, Englewood Cliffs, N.J., Prentice-Hall, 1972, xi, 132 pp.

Westing, J.H. and others. *Industrial Purchasing—Buying for industry and budgetary institution*, 2nd ed., N.Y., Wiley, 1956, x, 421 pp.

________, *Purchasing Management—Materials in motion*, 3rd ed., N.Y., Wiley, 1969, x, 538 pp.

Willets, W.E. *Fundamentals of Purchasing*, N.Y., Appleton-Century-Crofts, 1969, 189 pp.

C. Inventory Control

American Production and Inventory Control Society. *Proceedings of the 14th Annual International Conference of the ...* Washington, D.C., the Society, 1971, 360 pp.

Arrow, K.H., Karlins, S. and Scarf, H. *Studies in the Mathematical Theory of Inventory and Production*, Stanford, California, Stanford University, 1965, ix, 340 pp.

Baily, P. *Successful Stock Control by Manual Systems*, London, Gower Press, 1971, xi, 167 pp.

Baily, P. and Tavernier, G. *Design of Stock Control Systems and Records*, London, Gower Press, 1970, xi, 137 pp.

Barrett, D.A. *Automatic Inventory Control Techniques*, London, Business Books, 1969, viii, 156 pp.

Battersby, A. *Guide to Stock Control*, London, Pitman, 1962, v, 130 pp.

Bocchino, W.A. *New Tools and Techniques for Production and Inventory Control*, London, Industrial and Commercial Techniques, 1966.

British Institute of Management. *Control of Stock for Distribution—The practice of ten companies*, London, the Institute, 1955, 35 pp.

Brown, R.G. *Decision Rules for Inventory Management*, N.Y., Holt, Rinehart and Winston, 1967, x, 398 pp.

________, *Statistical Forecasting for Inventory Control*, N.Y., McGraw-Hill, 1959, xvii, 232 pp.

Buchan, J. and Koenigsberg, E. *Scientific Inventory Management*, New Delhi, Prentice-Hall of India, 1966, xiii, 523 pp.

Buffa, E.S. and Taubert, W.H. *Production-inventory Systems—Planning and control*, Homewood, Illinois, Irwin, 1972, xiv, 616 pp.

Burton, J.H. *Stores Accounts and Stores Control*, London, Pitman, 1956, v, 292 pp.

Cantor, J. *Profit-oriented Manufacturing Systems*, N.Y., American Management Association, 1969.

Caroll, P. *Practical Production and Inventory Control*, N.Y., McGraw-Hill, 1966, xiii, 338 pp.

Chadda, R.S. *Inventory Management in India*, New Delhi, Allied Publishers, 1964, ix, 118 pp.

Childs, G.L. *Unified Orders and Inventories*, North-Holland, 1967, 156 pp. ('Contributions to Economic Analysis', V. 48)

Compton, H.K. *Storehouse and Stockyard Management*, London, Business Books, 1970, viii, 491 pp.

D'anna, J.P. *Inventory and Profit—The balance of power in buying and selling*, N.Y., American Management Association, 1966, 123 pp.

De Garmo, E.P. *Materials and Process in Manufacturing*, 3rd ed., N.Y., Macmillan, 1969, 949 pp.

Dennett, H. *Unit Stock and Store Control*, London, Business Publications, 1963, 206 pp.

Ellon, S., Lampkin, W. *Inventory Control Abstracts* (1953–1965). London, Oliver & Boyd, 1968, 239 pp.

Enrick, N.L. *Inventory Management—Installation, operation and control*, Scranton, Pa., Chandler Publishing, 1968, 128 pp.

Entelek, Inc. *Programmed Instruction in Inventory Management*, Newbury Port, Massachusetts, 1964.

Fabrycky, W.J. *Procurement and Inventory Theory*, Oklahoma, State University Office of Engineering Research, 1964–65, 3 Vols. (Oklahoma State University Engineering Exper. 'Research Bulletin' Publication Nos., 140–141, 146)—V.1. 'Multisource item, concept'. V.2. 'Manufacture or purchase decision'; V. 3. 'Multi-item multi-source concept'.

__________, Banks, J. *Procurement and Inventory Systems—A theory and analysis*, N.Y., Reinhold, 1967, xii, 239 pp.

Fetter, R.B. and Dalleck, W.C. *Decision Models for Inventory Management*, Homewood, Illinois, Irwin, 1961.

Foster, D. Ed. *Automotive Warehouse*, London, Iliffe Books, 1970, 238 pp.

Fourre, J.P. *Applying Inventory Control Techniques*, N.Y., American Management Association, 1969.

Gopalakrishnan, P. *An Overview of Computers in India*.

Great Britain, Ministry of Technology. *Methods of Stock Control in the Carpet Industry*, London, the Ministry, 1966, iii, 36 pp.

Greene, J.H. *Production and Inventory Control Handbook*, McGraw-Hill, 1970, Various pages.

Hadley, G. and Whitin, T.M. *Analysis of Inventory Systems*, Englewood Cliffs, N.J., Prentice-Hall, 1963, xi, 452 pp.

Hill, G.V. and Smith, B.P. *Efficient Warehouse Operation*, London, Industrial and Commercial Techniques, 1972, 99 pp.

Hirsch, A.A. and Lovell, M.C. *Sales Anticipation and Inventory Behaviour*, N.Y., Wiley, 1969, 256 pp.

Hobbs, J.A. *Control Over Inventory and Production*, London, McGraw-Hill, 1973, x, 99 pp.

Hoffman, R.A. and Gunders, A. *Inventories—Control, costing, effect upon income and taxes*, 2nd ed., N.Y., Ronald, 1970, 444 pp.

Holt, C.C. and others. *Planning Production Inventories and Work Force*, Englewood Cliffs, N.J., Prentice-Hall, 1960, xii, 419 pp.

Jenkins, C.H. *Modern Warehouse Management*, N.Y., McGraw-Hill, 1968, ix, 319 pp.

Khare, B.D. *Inventory Control*, New Delhi, National Productivity Council, 1971, 46 pp.

Killeen, L.M. *Techniques of Inventory Management*, N.Y., American Management Association, 1969, 175 pp.

Krishnamurty, K. and Asstry, D.U. *Inventories in Indian Manufacturing*, Bombay, Academic, 1970, xx, 243 pp.

Lewis, C.D. *Scientific Inventory Control*, London, Butterworths, 1970, viii, 209 pp.

Lines, A.H. and Beart, J. *Inventory Control Techniques*, London, Industrial and Commerical Techniques, 1972, 143 pp.

Lirman, B.E. *How to Control and Reduce Inventory*, Englewood Clifs, N.J., Prentice-Hall, 1972, 197 pp.

Lockyer, K. *Stock Control—A practical approach*, London, Cassell, 1972.

Mack, R. *Information, Expectations, and Inventory Fluctuation—A study of materials stock on hand and on order*, N.Y., National Bureau of Economic Research, 1967, xii, 306 pp.

McGarrah, R.E. *Production and Logistics Management—Text and cases*, N.Y., Wiley, 1963, 560 pp.

McKibbin, B.N. *Practical Steps in Warehouse Design and Operation*, London, Industrial and Commercial Techniques, 1971, 185 pp.

Magee, J.F. *Production Planning and Inventory Control*, N.Y., McGraw-Hill, 1958, v, 333 pp.

__________, Boodman, D.M. *Production Planning and Inventory Control*, 2nd ed., N.Y., McGraw-Hill, 1967, x, 397 pp.

Mantri, M.K. *Stock Control*, Bomaby, Maharashtra State Road Transport Corporation, 1965, 73 pp.

Merchandising and Marketing Development Limited, London. *Gold in Your Hands—A guide to profitable stock management*, London, HMSO, 1966, top.

Moran, P.A.P. *Theory of Storage*, London, Methuen, 1959.

Morrell, A.J.F., Ed. *Problems of Stocks and Storage*, London, Olivera & Boyd (for Imperial Chemical Industry), 1967, viii, 61 pp. (Imperial Chemical Industry Monograph No. 4).

Morrison, A. *Storage and Control of Stock for Industry and Public Undertakings*, London, Pitman, 1969, xiv, 273 pp.

Morse, P.M. *Queues, Inventories and Maintenance*, N.Y., Wiley, 1965, ix, 202 pp.

Mulvihill, D.F. *Inventory Control Systems for Small Wholesalers*, Alaska, University of Business Research, 1963, 28 pp. (Small Business Administration Management Program).

Naddor, E. *Inventory Systems*, N.Y., Wiley, the Association, 1964, 115 pp.

National Foremen's Institute. *Purchasing Agent's Desk Manual*, Waterford, Connecticut, the Institute, 1968, 250 pp. (Looseleaf).

National Productivity Council. *Stores and Inventory Control in United States of America, Japan and West Germany*, New Delhi, the Council, 1961, 55 pp.

Niland, P. *Production Planning, Scheduling and Inventory Control—Text and cases*, London, Macmillan, 1970, xiv, 553 pp.

Palit, A.R., Ed. *Modern Trends in Purchasing, Store-keeping and Materials Management—A selection of articles from Vols. I and II of the Eastern Purchasing Journal*, New Delhi, Eastern Purchasing Journal Publications, 1962, 100 pp.

Plossl, G.W. and Wight, O.W. *Production and Inventory Control,* New Delhi, Prentice-Hall of India, 1979, xvi, 432 pp.

Prabhu, N.U. *Queues and Inventories—A study of their basic stockistic process,* N.Y., Wiley, 1965, 275 pp.

________, *Time-dependent Results in Storage Theory,* London, Methuen.

Prichard, J.W. and Eagle, R.H. *Modern Inventory Management,* N.Y., Wiley, 1965, 419 pp.

Reisman, A. and Others. *Industrial Inventory Control.* N.Y., Gordon & Breach Science Publishers, 1972, 180 pp. (Studies in Operations Research).

Robinson, O.P. and others. *Store Organization and Operation,* 2nd ed., Englewood Cliffs, N.J., Prentice-Hall, 1957, xv, 490 pp.

Sarkar, D.N. *Stock Accounting and Control,* Calcutta, Academic, 1969, 174 pp.

Scarf, H.F., Gilford, D.M., Shelly, M.W., Eds. *Multistage Inventory Models and Techniques,* Stanford Unviversity Press, 1963.

Stanback, T.M. *Postwar Cycles in Manufacturers' Inventories,* N.Y., National Bureau of Economic Research, 1962.

Starr, M.K. and Miller, D.W. *Inventory Control—Theory and practice,* New Delhi, Prentice-Hall of India, 1977, ix, 354 pp.

Stockton, R.S. *Basic Inventory Systems—Concepts and analysis,* Rockelight, N.J., Allyn, 1965, 109 pp.

Thomas, A.B. *Inventory Control in Production and Manufacturing,* Boston, Massachusetts, Cahners, 1969.

________, *Stock Control in Manufacturing Industries,* London, Gower Press, 1968, xiv, 213 pp.

Urwick, L.F. *Department Store—An informal talk on store organisation,* London, Urwick, Orr & Partners, 1960, 23 pp.

Vandemark, R.L. *Inventory Control Techniques,* 2nd ed., Dallas, Texas, Vandemark Inc., 1972, 429 pp.

________, *Revised Inventory Control Techniques,* Dalls, Texas, Vandemark Inc., 1965, 192 pp.

Van Hees, R.N. and Monhemius, W. *Introduction to Production and Inventory Control,* London, Macmillan, 1972, xiv, 146 pp.

________, *Production and Inventory Control—Theory and practice,* London, Macmillan, 1972, xvi, 370 pp.

Wagner, H.M. *Statistical Management of Inventory Systems,* N.Y., Wiley, 1962.

Warman, J. *Warehouse Management.* N.Y., Heinemann, 1971, 194 pp.

Welch, W.E. *Tested Scientific Inventory Control,* Greenwich, Management Publishing Co., 1972.

Whitin, T.M. *Theory of Inventory Management,* 2nd ed., Princeton, Princeton University Press, 1967.

Wolf, T.R. *Better Management of Maintenance Stores and Inventories,* London, Industrial and Commercial Techniques, 1971, 45 pp.

II. PERIODICALS

Apics Annual Conference Proceedings. *American Production and Inventory Control Society*, International Headquarters, Suite 504, Watergate Building, 2600 Virginia Avenue, NW, Washington D.C. 20037.

Chemical Purchasing. Myers Publishing Co., Inc., 381 Park Avenue, S., New York, N.Y. 10016.

Indian Buyer. Short Street, Calcutta-17.

Industrial Purchasing Agent. Publications for Industry, 21 Russell Woods Road, Great Neck, N.Y. 11021.

Journal of Purchasing. National Association of Purchasing Agents, Inc., 11 Park Place, New York, N.Y. 10007.

Materials Management Journal of India—D–409, Defence Colony, New Delhi-3.

Michigan Purchasing Management. Ricker Publishing Co., 504, Blvd. Center, Building, Detroit, Michigan 48202.

Modern Purchasing. Maclean-Hunter Publishing Co., 481, University Avenue, Toronto 2, Canada.

Production and Inventory Management. American Production and Inventory Control Society, 2600 Virginia Avenue, Washington, D.C. 20039.

Purchasing. 111 Allied Industrial Estate, MM Chotani Road, Bomaby-16.

Purchasing—The methods and news magazine for Industrial buyers. Conover-Mast Publications, Inc., 205E, 42nd Street, New York, N.Y. 10017.

Purchasing Bulletin. Institute of Purchasing and Supply, York House, Westminster Bridge Road, London El. England.

Purchasing Journal. Purchasing Officers Association, York House, Westminster Bridge Road, London El. England.

Purchasing Week. McGraw-Hill Inc., 330W., 42nd Street, New York, N.Y. 10036.

Index